MATTERING PRESS

Mattering Press is an academic-led Open Access publisher that operates on a not-for-profit basis as a UK registered charity. It is committed to developing new publishing models that can widen the constituency of academic knowledge and provide authors with significant levels of support and feedback. All books are available to download for free or to purchase as hard copies. More at matteringpress.org.

The Press' work has been supported by: Centre for Invention and Social Process (Goldsmiths, University of London), European Association for the Study of Science and Technology, Hybrid Publishing Lab, infostreams, Institute for Social Futures (Lancaster University), OpenAIRE, Open Humanities Press, and Tetragon, as well as many other institutions and individuals that have supported individual book projects, both financially and in kind.

We are indebted to the ScholarLed community of Open Access, scholar-led publishers for their companionship and extend a special thanks to the Directory of Open Access Books and Project MUSE for cataloguing our titles.

MAKING THIS BOOK

Books contain multitudes. Mattering Press is keen to render more visible the unseen processes that go into the production of books. We would like to thank Anna Dowrick and Uli Beisel, who acted as the Press' coordinating editors for this book, Joe Deville for his additional editorial input, Jennifer Horsley for her work on the book production, Steven Lovatt for the copy editing, Melanie Mallon for the formatting, Alex Billington and Tetragon for the typesetting, and Will Roscoe, Ed Akerboom, and infostreams for their contributions to the html versions of this book.

COVER

Cover art by Tabita Rezaire; cover design by Julien McHardy

What happens if we sincerely examine the global impact of technology from the bottom up? We get a taste of what that might look like in this rich and thrilling collection of essays. Individually, the essays offer distinct case studies, and together they challenge us to cement the identity of the Global South for what it has long been: a site of inventiveness and originality when it comes to technology. The connections traced in this book span regions and nations not normally associated with technological innovation, and they lead us to set aside, indeed redefine, our long-held beliefs about the impact of technology on different societies.

TECHNOSCIENTIFIC GLOBALISATION FROM BELOW

EDITED BY

MATHIEU QUET

KOICHI KAMEDA

JESSICA POURRAZ

AND

YVES-MARIE RAULT-CHODANKAR

Mattering Press

First edition published by Mattering Press, Manchester.

Cover image: 'Dilo'. 2017. Courtesy of Tabita Rezaire and Goodman Gallery
Cover design by Julien McHardy

Freely available online at https://www.matteringpress.org/books/
technoscientific-globalisation-from-below

ISBN: 978-1-912729-38-8 (pbk)

ISBN: 978-1-912729-36-4 (pdf)

ISBN: 978-1-912729-35-7 (epub)

ISBN: 978-1-912729-37-1 (html)

DOI: http://doi.org/10.28938/mw20-c7f3

Mattering Press has made every effort to contact copyright holders and will be glad to rectify, in future editions, any errors or omissions brought to our notice

CONTENTS

Figures   9

Tables   11

Funding acknowledgements   12

Author biographies   13

1 · Introduction. Rethinking technoscientific globalisation with the Global
    South   17
        The Technoglob Collective

SECTION 1. LIVING OFF THE INFORMAL

2 · Enabling and resisting the platform economy from below:
    Platform immigrant workers in Ecuador   57
        Henry Chávez and María Belén Albornoz

3 · Digital knowledge from below: Low-skilled labour migration
    to the Gulf countries and technology adoption in India   81
        Javed Mohammad Alam

SECTION 2. NAVIGATING INTERNATIONAL INEQUITIES

4 · Calibrating the global: How are Ghanaian scientists shifting
    Africa's position in global atmospheric science?   113
        Jessica Pourraz and Allison Felix Hughes

5 · Affirming pharmaceutical sovereignty: Technology transfer agreements
    and vaccine geopolitics during a global health emergency   140
        Koichi Kameda, Denise Pimenta, and Gustavo Matta

6 · A human drug amid animal diseases: The ecology of
globalised heparin 174
Thibaut Serviant-Fine

SECTION 3. ADJUSTING THE GLOBAL

7 · Patching development: Information technology adjustments
in the Mauritian logistical sector 199
Marine Al Dahdah and Mathieu Quet

8 · Halting the 'forced march': The ups and downs of Chad's
integration into global pharmaceutical markets 221
Ilyass Mahamat Nour Moussa

SECTION 4: CREATING ALTERNATIVE VALUES

9 · Making value off-patent: India's pharmaceutical globalisation 247
Yves-Marie Rault-Chodankar

10 · The division of biometric labour: Relations of production
in African voter-identification technologies 272
Cecilia Passanti

11 · How magic bullets travel: An account of ready-to-use
therapeutic food in India 298
Aamod Utpal

Photos of the Technoglob collective 328

FIGURES

FIG. 2.1 Platform workers looking at their smartphones for orders to deliver in Quito, Ecuador 57

FIG. 2.2 Platform drivers detained by the police near Quito airport during social protests in Quito, Ecuador 72

FIG. 3.1 Indian labourers at Shamkha 19, Abu Dhabi 82

FIG. 4.1 Light-scattering techniques to measure particulate matter (PM) 122

FIG. 4.2A PurpleAir sensor 125

FIG. 4.2B Clarity Node sensor 125

FIG. 4.2C ZeFan sensor 126

FIG. 4.2D Modulair sensor 126

FIG. 4.3 Plantower sensor 127

FIG. 4.4A Collocation at the University of Ghana of LCS with fixed reference-grade monitors 129

FIG. 4.4B Calibration process, Accra, Ghana 129

FIG. 4.5A Openaq low-cost sensor kit 133

FIG. 4.5B Openaq printed control board 133

FIG. 4.5C Openaq sensor 134

FIG. 5.1 'Vacinas para todos já!!!' / 'Vaccines for all now!!!', Teresópolis, Rio de Janeiro 140

FIG. 7.1 Along the Mauritius Free Port Zone 200

FIG. 8.1 National coordination of the SWEDD project, N'Djamena 230

FIG. 8.2 Overview of the building housing the medicines quality control laboratory, N'Djamena 233

FIG. 9.1 Product display in three corporate offices 255

FIG. 9.2 A cGMP-compliant factory, inaugurated in 2019, and a factory being constructed, Pune 259

FIG. 9.3 Some visiting cards collected during fieldwork 262

FIG. 10.1 The system administrator of the biometric citizen database of Senegal 278

FIG. 10.2 Civic education video clip depicting two officials biometrically verifying the identity of a voter at a polling station, Nairobi, Kenya, May 2017 279

FIG. 10.3 A generic biometric identification system 280

FIG. 10.4 The division of biometrics labour between African public administration and the biometric market 292

FIG. 12.1 – 12.12 Glimpses of the technoglob collective 328

TABLES

TABLE 1 Comparison of value-making practices in the pharmaceutical
industry   251

TABLE 2 Number of UNICEF suppliers of RUTF and their locations,
2023 (UNICEF RUTF market update 2023)   305

FUNDING ACKNOWLEDGEMENTS

The publication of this book has been funded by a grant from the French Agence Nationale de la Recherche (ANR) (Anipharm, 19-CE26-0001), with additional funding from the Institut Francilien Recherche Innovation Sociétés (IFRIS) and the Global Research Institute of Paris (GRIP).

AUTHOR BIOGRAPHIES

JAVED MOHAMMAD ALAM is a PhD candidate in STS at the Centre Population et Développement (CEPED), Université Paris Cité. He studies the role of digital technologies in the making of contemporary financial inclusion policies in India.

MARINE AL DAHDAH is a CNRS researcher at the Centre d'étude des mouvements sociaux (CEMS-EHESS) and a member of Unit 1276 'Risks, Violence, Reparation' of the French National Health and Medical Research Institute (INSERM). She is an associate researcher at the French Institute in Pondicherry and the Center for Human Sciences (CSH) in Delhi (India). Her research focuses on health policies in Asia and Africa, and more particularly on digital healthcare in India, Ghana, and Kenya. She is the author of *Mobile (for) Development: When Digital Giants Take Care of Poor Women* (Cambridge University Press 2022).

MARÍA BELÉN ALBORNOZ is a professor and researcher at the Latin American Faculty of Social Sciences (FLACSO-Ecuador). She is the principal investigator of the Fairwork Project in Ecuador. Her current research concerns technology transfer models; public policy of science, technology, and innovation; big data; ICTs in education; social innovation; and fairwork. In 2017 she was a Fulbright Scholar at Rensselaer Polytechnic Institute and a Visiting Scholar at Aalborg University in Denmark, where she worked on theorising imaginaries of innovation and policy network analysis.

HENRY CHÁVEZ is an associate researcher at CEPED (Université Paris Cité), the Institut de Recherche pour le Développement (IRD) and CTS Lab FLACSO Ecuador. He holds a PhD in social sciences from the École de hautes études en sciences sociales (EHESS) in Paris and has an interdisciplinary background in social sciences, economics, politics, and data science. He has worked as a researcher and consultant in the public and private sectors, as well as in social organisations, NGOs, and international agencies. His work focuses on socio-economic cycles, techno-economic transformations, social studies of science and technology (STS), big data artificial intelligence, and platform economics.

ALLISON FELIX HUGHES is a senior lecturer at the Department of Physics, University of Ghana. He was also a Special Fellow at the Faculty of Arts and Science of Harvard University, USA. His research interests include air pollution characterisation and modelling, renewable energy technologies, climate change, physics education, and conversion of municipal solid waste (MSW) to energy. He is a member of the Ghana Institute of Physics; Ghana Science Association; Institute of Physics, UK; and the University Teachers Association of Ghana.

KOICHI KAMEDA DE FIGUEIREDO CARVALHO holds a PhD in sociology (EHESS, Paris) and a BA in law (UERJ, Brazil). He is a postdoctoral researcher at the Center Emile Durkheim at the University of Bordeaux and is an associate researcher at the Interdisciplinary Center for Public Health Emergencies (NIESP/Fiocruz, Brazil) and at CEPED/Université Paris Cité. His work focuses on issues related to biomedicine, global health, intellectual property, and pharmaceutical markets and social justice, and he has worked on local manufacturing and regulation of diagnostics and vaccines in the Global South and public health–oriented R&D models.

GUSTAVO MATTA holds a PhD in public health from the Institute of Social Medicine at the State University of Rio de Janeiro. He is currently a senior researcher in Public Health at the Data and Knowledge Integration Centre for Health (CIDACS) at the Oswaldo Cruz Foundation in Bahia (FIOCRUZ). In addition, he serves as coordinator of the Interdisciplinary Public Health Emergencies Unit (NIESP/CEE/FIOCRUZ) and the Zika Social Sciences Network at Fiocruz and is a member of the Fiocruz Graduate Program for Internationalisation. His work particularly addresses public health emergencies and re-emergencies, primary health care, vaccine politics, health policies, and global health perspectives from the Global South.

Ilyass Mahamat Nour Moussa is a PhD candidate in STS at the CEPED, Université Paris Cité. His research focuses on the construction of the Chadian pharmaceutical market in relation to informality.

CECILIA PASSANTI holds a PhD in science, technology, and society (STS) at Université Paris Cité and is an associate researcher at the Center for Population and Development (CEPED). Her thesis, titled 'Les infrastructures numériques du vote en Afrique. Biométrie, machines à voter et marchands de démocratie au Kenya et au Sénégal', is a study of election technologies and biometrics in Africa. Using ethnographic, historical, and sociological methods, she explores new forms of international and industrial governance of public participation and citizenship. She has recently published 'The (Un)making of Electoral Transparency through Technology: The 2017 Kenyan Presidential Election Controversy' (*Social Studies of Science* 2022), and 'Contesting the Electoral Register during the 2019 Elections in Senegal. Why Allegations of Fraud Did Not End with the Introduction of Biometrics' (Francia 2021).

DENISE PIMENTA is a social anthropologist (PhD, University of Sao Paolo) and currently a postdoctoral fellow at the Center for Data and Knowledge Integration for Health at the Oswaldo Cruz Foundation (Cidacs-Fiocruz/BA), where she is a member of the Public Engagement with Science team. She is also a researcher at the Interdisciplinary Center for Public Health Emergencies (NIESP-Fiocruz). During her PhD, she did fieldwork in Sierra Leone (West Africa) to understand the relationship between the Ebola epidemic and gender issues.

JESSICA POURRAZ holds a PhD in sociology (2019) from the École des hautes etudes en sciences sociales (EHESS) in Paris. She is a postdoctoral research fellow with Sciences Po Bordeaux and an associate researcher at the Center for Human Sciences (CSH) in Delhi. Her research focuses on issues related to science, biomedicine, the environment, and health, and more particularly on the health effects of air pollution in India and Ghana.

MATHIEU QUET, sociologist, is a research director at Institut de recherche pour le développement, Paris, and a member of the Centre Population et Développement (CEPED), Université Paris Cité. His research focuses upon the social aspects of pharmaceutical development in India, and he is more generally interested in observing the globalisation of technological markets from Global South countries. He authored *Illicit Medicines in the Global South. Public Health Access and Pharmaceutical Regulation* (Routledge 2021).

YVES-MARIE RAULT-CHODANKAR is an associate professor at Université Paris 1 Panthéon-Sorbonne and an associate researcher at the Pôle de recherche pour l'organisation et la diffusion de l'information géographique (Prodig). His work sits at the intersection of economic geography and development studies, examining how emerging economies integrate into global value chains and drive regional development in the Global South. Through extensive fieldwork in India, Yves-Marie's ethnographic research captures the lived realities of local entrepreneurs, particularly in the pharmaceutical sector, to show how they adapt and innovate to reshape global production networks from below.

THIBAUT SERVIANT-FINE trained as a pharmacist and historian of medicine. He completed a postdoc at the Centre Population et Développement (CEPED) in Paris, researching the pharmaceutical uses of animal life, and subsequently opened a science café.

AAMOD UTPAL is a PhD candidate in STS at the Centre Population et Développement (CEPED), Université Paris Cité. His research focuses on analysing contemporary public health topics using STS approaches. He is currently working on the circulation of a humanitarian technology for severe acute malnutrition – ready-to-use therapeutic food – in India.

I

INTRODUCTION. RETHINKING TECHNOSCIENTIFIC GLOBALISATION WITH THE GLOBAL SOUTH

The Technoglob Collective (Javed Mohammad Alam, Henry Chávez, Mahamat Nour Moussa Ilyass, Koichi Kameda, Cecilia Passanti, Jessica Pourraz, Mathieu Quet, Yves-Marie Rault-Chodankar, Aamod Utpal, with Mariana Gameiro and Thibaut Serviant-Fine)

IN LATE 2023, THE HOUTHI MOVEMENT TOOK THE WORLD BY SURPRISE when it launched a series of drone attacks against ships navigating through the Bab-el-Mandeb strait, in support of the Palestinian cause. This Islamist group, despite leading a violent uprising against Yemen's government since 2004 and controlling a large part of the country, was barely known beyond the Middle East before this event. The Supporters of God (Ansar Allah), as they call themselves, organised in the mountains of northern Yemen during the 1990s, under the leadership of a local religious figure. The airborne attacks they carried out demonstrated their capacity to alter the march of the world: while they were threatening the strait, four of the world's largest shipping companies stopped using it, diminishing global trade flows by 12% according to some analysts. These attacks challenged the prevailing notion that global dynamics are solely influenced by major technological players – the richest nations and multinational corporations. Targeting the lifelines of modern economic globalisation – super-tankers, megaships – they prompted global superpowers like the United States, India, and France to collaborate in deploying planes, warships, and drones to protect shared economic interests.

The rapidity and intensity of this international response accompanied a reassessment of the image of Houthis, from perceived uneducated, technology-averse militants to sophisticated military actors on the global stage. However, within certain online communities outside the Western world, the movement's technological prowess and military endeavours were already recognised and sometimes celebrated. On the Chinese-owned video application TikTok, thousands followed Rashid Al Hadad, a 19-year-old Yemenite who posted videos of himself on the *Galaxy Leader*, a massive Japanese ship captured on the Red Sea. Furthermore, the Houthis kept on displaying their acute awareness of the strategic importance of technological infrastructure. In February 2024, on the encrypted telecommunication platform Telegram, the group released a map detailing submarine cable routes in the Red Sea, through which approximately 17% of global internet traffic transits. Their threat to destroy this infrastructure appeared as the possibility of a serious blow to Europe-Asia technological communication.

Although at that time the Houthis did not own submarines able to dive deep enough and locate the cables in the Red Sea, the risk they represented could not be dismissed. In fact, they had displayed a great ability to harness the technological inventiveness of Global South countries and populations, despite their exclusion from mainstream international relationships.[1] The HESA Shahed 136 drones, which had been used by the Houthis to attack boats in the Red Sea, originated from the Iranian military-industrial complex: they had been designed by two companies which incorporated a gasoline jet engine, an uncommon feature for drones, usually powered by Lithium-ion batteries. According to specialists, they were the cheapest drones in the world, priced at around 20,000 dollars, a bagatelle in comparison to the American and European missiles, which can cost 2 million dollars per piece (Sabbagh 2024). On the 9th of January, Britain's HMS *Diamond* intercepted 7 of the 18 drones launched by the Houthis, notably utilising Sea Viper and Aster missiles, which cost between 1 and 2 million dollars each. The low cost of the Iranian drones largely owed to the incorporation of technologies made in BRICS countries, such as Chinese voltage converters and Russian navigation systems, which were in use on the Russian-Ukrainian battleground at the same period. Indeed, the Russian government had collaborated with the Iranian authorities to develop and manufacture its own version of the Shahed 136 drone, the Geran-2, which Ukrainian soldiers sarcastically called the 'mopeds' or 'lawnmowers' because of the loud sound emitted by their engines.

The Houthis' drones, designed and produced in the Global South, signal the emergence of new geopolitical alignments that challenge the idea of global order and the military hegemony of countries hosting the highest capital-intensive technologies. But they also illustrate the capacity of Global South countries and populations to develop original innovations despite limited financial capacity and, in this extreme scenario, exclusion from dominant production networks. This situation also underscores the necessity of broadening our perspective to include oft-ignored social groups in the analysis of the production, consumption, and global circulation of technologies. It shows that the forces simultaneously shaping the future of global technoscience and international trade can emerge from the most unexpected places, challenging conventional wisdom and prompting a reassessment of global power processes. The Global South countries host some of the key dynamics that shape global technological capitalism, and this realisation is pivotal for understanding the complex, multifaceted nature of technoscientific globalisation. Looking at technoscientific globalisation 'from below' – that is, from non-hegemonic countries and subalternised groups – provides a critical lens through which to reexamine the interplay between technology, power, and society at the global scale. Our edited volume builds on such an effort.

GLOBAL SOUTH CONTRIBUTIONS TO TECHNOSCIENTIFIC GLOBALISATION

Be it through technological uses, production, or innovation processes, Global South populations, firms, and governments play a key role in shaping the relations between sciences, technologies, and markets at an international scale. This phenomenon is even more striking as Global South countries have long been considered as cut off from globalising dynamics, as exclusive extractive sites, or as remote recipients for technical innovations – last in line and benefitting from them only a long time after the richest countries. Although historians of science and science studies scholars have demonstrated the geographical variety of knowledge streams fashioning global science and technology, as stated in Joseph Needham's metaphor[2] (Needham 1964), dominant discourse has kept 'othering' the South, implicitly considering it foreign to rationality, science, and innovation (Prasad 2022; Said 1997). But attention to the actual developments of global capitalism makes such a preconception even more erroneous. Across most technical domains, the globalisation of technological innovations increasingly affects and shapes Global South countries, not only as sites of extraction, but also as sites of production, innovation, and consumption. In return, Global South populations increasingly participate in the fabrication of technologies, through innovation and production practices, as users adjusting and diverting technological goods from their initially intended uses, or as locations for the dismantling and relegation of unused technical objects, from massive container ships to TV screens. As a result, we can observe the contribution of the Global South to technological globalisation in multiple domains.

Innovation – First, while dominant discourse tends to frame the Global South as the cradle of copy industries, unable to come up with radical innovation, recent dynamics suggest that this vision might be skewed. Although the fact is concealed by the power structures of the academic field, Sub-Saharan African universities, for instance, play an increasing role in global knowledge production (Gebremariam et al. 2023). In the IT sector, through developing platform technologies or geospatial tools adapted to local needs, many African states, from Rwanda to South Africa, have notably elaborated forms of government at the national or city scale that are 'smart' – digitally tooled – in their own ways (Odendaal 2023) and despite the persistence of multiple forms of digital marginalisation (Chacko 2022). In the medical field, the Global South countries have been developing treatments sometimes more efficient than the products of capital-intensive R&D firms from the North, as illustrated by the dramatic contribution of Nepalese and Indian ophthalmologists to eye surgical science and to the low-cost and rapid treatment of cataract problems (Williams 2019). In the agricultural sector, attention to long-term solutions has led to the questioning of supposed universal technological fixes and the elaboration of new scientific paradigms, as illustrated by alternative treatments for Panama disease in Filipino banana plantations (Paredes 2023). This ability to locally develop systems of innovation and infrastructures, not only depending on imported technologies but benefitting from local engineering knowledge and skills, is nothing new – and scholarship shows that even during the colonial period, local engineers and inventors contributed to the establishment of massive infrastructural projects, as in the case of India's electricity supply system in West Bengal (Sarkar 2021). However, this ability of Global South countries to innovate is now becoming more important in defining what technical innovation is and should be, at an international scale, and in some fields this innovative capacity and the variety of paths it opens is particularly visible: nowhere in the world are there more innovative processes around energy supply than in subtropical regions, where alternative energy procurement systems are frequently harnessed through off-grid systems (Boamah 2020). In that regard, Global South countries undeniably play a growing part in the expansion of global markets for technologies through multiple forms of creation and innovation.

Production – Additionally, the Global South hosts critical sites for the industrial production of technological devices, as illustrated by the fact that most ICT devices are exported from Asia, with local companies producing their own smartphone models (Chinese Oppo, Huawei, Xiaomi, Indian Micromax), along with the models from US brands such as Apple's smartphones, produced by Taiwanese firm Foxconn across Asian countries. Global South countries tend to differ from the richest ones in often taking up the production process of technologies, since Northern countries keep on outsourcing growing shares of industrial production to Southern firms. India's pharmaceutical industry stands out in supplying medicines to the world. For example, most antiretroviral medicines used against HIV in the developing world come from India (Rajan 2017). Brazil, after the USA, is the second largest producer of genetically modified (GM) crops (Aga 2021; Peschard

2022), and emerging countries do not depend only on crops that are selected in the Global North, sometimes developing their own seed banks through alternative means (Chacko 2022).

Extraction – Furthermore, even though many Global South countries are still too often considered as massive extraction sites by governments and companies alike (Gudynas 2020; Tsing 2005), they no longer extract resources only for others' benefit. The 'resource curse' that has long plagued developing countries, rendering them unable to draw development benefits from mineral-rich ground, is being challenged and reconfigured. Mobile phone batteries across the world still widely rely on the labour of Congolese miners and traders (Smith 2021), and the nuclear plants of the Global North are filled with Nigerien uranium (Hecht 2014). But at the same time, Southern governments and populations are increasingly aware of the power they hold over the world's technological futures, as illustrated by the intensity of environmental struggles in Peruvian Amazonia (Buu-Sao 2019). This is also true for biological extraction: in the context of the global fight against influenza, Southern countries have been particular places of extraction with no assurance of benefit sharing or accountability, but this is beginning to be contested. When the World Health Organisation's labs transferred Indonesian H5N1 samples to a pharmaceutical company without consulting Indonesia, the country withdrew from the WHO virus-sharing mechanism (Stephenson 2011). Moreover, what is extracted in the South can also be consumed locally. In Ghana, nuclear engineers hope that they will soon be able to put African uranium to good use within their own nuclear plants to generate energy (Osseo-Asare 2019).

Consumption – The Global South countries also represent growing markets for the consumption of multiple industrial and technological products and services. Cement has been adopted across Africa to the extent that local populations have developed an effective relationship with this material, which is used to materialise dreams of comfortable living spaces (Choplin 2023). The Sahel countries have become major users of surveillance technologies to control migration flows (Donko 2022). Financial technologies across Asian and African countries have been adopted swiftly, and populations have harnessed the possibilities provided by digital finance to facilitate payments, exchange money, and acquire loans (Donovan and Park 2022; Kusimba 2021). New technologies designed for cultural industries have met with great success in the developing world, the global expansion of platform economy infrastructures being a case in point (Boczkowski 2021; Bouquillion et al. 2023; Lobato 2019).

Relegation and revaluation – Global South countries' contribution to the globalisation of technologies is not restricted to innovation, production, extraction, and consumption. An important aspect of their participation in global capitalism consists in the handling and revaluation of relegated goods, waste, and residues. This can be illustrated through the case of second-hand and wrecked vehicles sent from rich countries to poorer ones. Indeed, African countries absorb 40% of the global flow of second-hand vehicles. This is not a new trend, but it is taking on a new dimension because of the more stringent regulations that Global North countries are imposing on their vehicle fleets. Growing awareness of the health effects of particulate matter (PM) is leading to a technological, environmental, and health imbalance to the detriment of countries with fewer control resources, to which fleets of relegated vehicles are directed. In this way, by reinforcing environmental regulation, the Global North countries are accelerating the obsolescence of their car fleets and fuelling the flow of exports to the countries of the Global South. Second-hand vehicle markets in Ghana, for instance, largely trade in obsolete vehicles coming from North America. The conditions of existence and transformation of these vehicles are embedded in the practices of various stakeholders, traders, and technicians, such as the so-called fitters (Powell 1995). This is the term for informal sector mechanics who repair vehicles in Ghana. Fitters are illegitimate in the eyes of the authorities, despite the fact they are involved in a process of value creation and innovation through maintenance and repair (Edgerton 2011). This situation also shows how materials can move from one category to another following different value sets and different contexts – for example, from waste to consumption goods, from matter extracted from animals to medicines (Appadurai 1996).

APPROACHING TECHNOSCIENTIFIC MARKETS 'FROM BELOW'

If Global South countries are fully participating in the expansion of global technoscience markets – whether as innovators, manufacturers, consumers, raw material providers, or scrap recyclers – then this participation needs to be thoroughly characterised. This book focuses primarily on how such activities manifest within Global South countries themselves, rather than on how these activities transform science and technology globally. But is there a distinct way in which technoscientific globalisation manifests in the Global South, and is it meaningful to consider the 'Global South' as a relevant entity? We argue that in both cases the answer is 'yes'. However, addressing these questions requires careful consideration and critical use of analytical tools, such as the concept of 'from below', which has guided the authors' work. The following section elaborates on this perspective and explains our approach in studying technoscientific globalisation 'from below'.

On one hand, the globalisation of technoscientific markets is not a uniform process; it unfolds in contrasting ways depending on where it takes place. On the other hand, regularities characterise shared features depending on geographical, historical, social, and economic variables. In that perspective, we contend that it makes sense to account for the experience of technoscientific globalisation shared by countries usually gathered under the loose category of 'Global South' – alternatively 'developing countries', formerly 'Third World,' and in some cases 'least developed countries' – all of them composing the 'Majority World' (Alam 2008). One major reason to highlight these shared experiences is to expose some of the structural economic inequalities and power asymmetries between countries that determine how technoscientific globalisation is experienced. Indeed, we think there is something obscene in the 'flat world' narrative where atomised entities would coexist through horizontal relations, and in the acritical celebration of both economic globalisation and technological solutionism. The observation of daily life and socio-anthropological fieldwork in (financially) resource-poor contexts often confronts the observer with radically different life experiences to their own, in which the exercise of human rights to life, to liberty, to security is much more constrained than in richer settings. The social sciences have a duty to account for such a difference. However, we condition the use of that broad category to empirical investigation, which allows us 1) not to reduce the experience of social actors in the Global South to one of exclusive subordination, and 2) to keep a constant focus on the singularity of these social actors that the 'Global South' category tends to obliterate. Furthermore, recent works have shown that such a category as 'Global South' needs to be diversified, understood as a 'polyphonic' category (Waisbich et al. 2021). 'South' and 'North' have become plural categories that now encourage multiple forms of analysis, from studies of the South in the North to ones of the North as seen from the South, and the East/West countries do not always fit unquestionably in either North or South. The category 'Global South', then, mostly follows the use of the 'South' as a category when it is advocated by 'the nations and groups of the South to assert their difference by means of scientific criticism' (Dumoulin Kervran et al. 2018).

From this position, the 'North-to-South' narrative of technological transfer, knowledge dissemination, or market fabrication is being challenged by the emergence of research and development capabilities across the globe and more specifically in countries of the Global South. No innovation process is fully local nor global; such processes are the product of an assemblage of knowledge and know-how from multiple origins (Bhaduri 2016). These dynamics of polycentric innovation, coupled with market strategies, often take original and discrete paths, as they are embedded in sets of rules that challenge the dominant forms of organisation of global capitalism (Arnold 2019; Quet 2022) and global science determined by the richest countries. As a result, emancipation and constraint become rearticulated in unexpected ways in the Global South considered as a site of innovation, production, and consumption. While new markets and promises are appearing (Donovan and Park 2022), emerging players such as India and China are disrupting the traditional geopolitical balance and hierarchies of innovation (Kaplinsky et al. 2009), reframing, for instance, the dominant conceptions of digital and pharmaceutical capitalism (Lei 2023; Lindtner 2020).

Global studies offers opportunities to analyse this phenomenon through the coinage 'globalisation from below'. In recent years, the social sciences have increasingly stressed the need to better integrate actors and objects 'from below' – often located in the Global South – into the analysis of economic globalisation (Choplin and Pliez 2018; Mathews et al. 2012; Tarrius 2002). A growing body of research has documented the role of subaltern groups in the dynamics of global change and market fabrication. This research has shown that the notion of 'globalisation from below' (Mathews et al. 2012) serves as a useful entry point to highlight issues often neglected by dominant analyses of economic globalisation. Non-hegemonic actors such as informal female workers (Gago 2017), migrants, marginalised groups and communities (Tsing 2015), and small intermediaries (Tastevin 2012) play a key role in the contemporary circulation of goods, people, and ideas, alongside more powerful and better-known actors. Engineers and scientists from the Global South have been fully part of this connective process even when their role has been erased or marginalised (Laveaga 2018).

The category of 'from below' has also been adopted by some STS scholars. In that sense, Sandra Harding (2008) argues that proposals to investigate modernity and its sciences ought to look at positions 'from below', which involves taking into consideration the standpoints of women and the world's other least-advantaged citizens. Likewise, recent STS studies have turned their attention to innovation 'from below' to consider the position of those with low socioeconomic and geopolitical status in global science. The qualifier 'from below' refers to people who are 'marginalised in some relationship of power', including in science and technology production (Williams 2019), though the choice of a perspective 'from below' does not preclude drawing on related concepts.

The notion of 'from below' remains fuzzy, however, as it encompasses a wide range of heterogeneous operations and actors who can be more or less dominated. The study of things and processes 'from below' – be they globalisation, markets, capitalism, history – is not new and has already fostered multiple debates. For instance, historians have provocatively asked 'who is below?'. 'Below' may vary according to the boundaries defining the category: for instance, the working class or the poor, if adopting socioeconomic criteria; the developing nations' populations, according to the critique of international trade; the women in patriarchal societies; the marginal, persecuted, or nonconformist groups and individuals, following social studies of deviance; or simply those not belonging to the elites (Cerutti 2015; Hailwood and Waddell 2013). In that regard, choosing to investigate 'from below' requires a serious effort at defining what kind of 'below' is at stake. Following a more general understanding, tracing history from below means restituting the 'roads that have not been travelled and that have lost the battle for legitimacy' (Cerutti 2015).

Recognising and building on the abovementioned literature that adopts a perspective 'from below', we propose in this book to make cautious use of that expression. We acknowledge its relevance, as it has oriented many of our discussions since we started working as a collective, but we also refuse to take the meanings of 'from above' and 'from below' for granted. This has two implications for the way in which we analyse technoscience and market expansion 'from below' throughout the book. Firstly, we insist on the power asymmetries and hierarchical differences implied by such expressions as 'from above' and 'from below' (Choplin and Pliez 2015). In that sense, to analyse 'from below' requires both approaching and defining 'the actors from above' as a contrasting entity. Secondly, we understand 'from below' as a broad category and acknowledge that the uses we make of the expression differ at times, depending on the phenomena we describe in the chapters. Though obviously encompassing the wide spectrum of vulnerable, invisible, and usually forgotten actors, it may also include others who are not 'subjugated' and less obviously identified as 'from below', such as elite scientists, experts and managers in the Global South, or emerging nations such as South Africa, India, and Brazil that share a 'betwixt-and-between status' (Pollock 2019). These are 'non-hegemonic' countries and actors in the sense of being dominated in the international division of scientific labour, though still building up room for manoeuvre to develop national policy (Losego and Arvanitis 2008) and pursue nation-building projects. Moreover, they might also establish new power inequalities and hierarchies vis-à-vis other countries and populations within the South. However, they all share 'vantage points from a peripheral space of global science' that becomes a 'location of insight' (Pollock 2019). As a result, our perspective on technoscience and market expansion in the Global South is as much 'from below' as we can make it, but every time this is required, we will clarify which 'below' we are dealing with.

ENQUIRING INTO GLOBAL SOUTH TECHNOSCIENTIFIC GLOBALISATION: AN ANALYTICAL FRAMEWORK

Over the last quarter of a century, STS has increasingly focused on non-Western countries. Pioneering works published in the 1990s laid the foundation for increasing scholarship in the 2000s (Abraham 1998; De Laet and Mol 2000; Harding 2008; Waast 1995). Special issues (Postcolonial Studies 2009; Science as Culture 2005; Science, Technology and Human Values 2014, 2016; Social Studies of Science 2002), handbook chapters (Anderson and Adams 2008), and dedicated handbooks (Harding 2011; Medina et al. 2014) gradually established 'postcolonial studies of technoscience' as a dynamic subset of STS. Research has shed light on the worldwide expansion of technoscience by examining it from the perspective of Global South countries. 'Going South' (Dumoulin Kervran et al. 2018) has enabled us to unpack the specific forms of dispossession or capital accumulation at play (Peterson 2014), the coexistence of contradictory disease ontologies and care practices (Langwick 2011), the adjustment of innovative practices to resource-scarce settings (Chee 2021), and the beliefs embedded in development or technological narratives (Invernizzi et al. 2008; Rottenburg 2009) – to mention but a few examples. This body of exciting scholarship has helped the STS community recognise the significant insights gained from studying technoscience beyond the wealthiest, most powerful, and academically prestigious sites (Pollock 2019). It has also firmly established that the technoscientific development of global societies is deeply intertwined with the countries of the Global South. In a way, extending Achille Mbembe's discussion, the world's technoscientific becoming is African, Asian, Latin American – it goes South (Mbembe 2020).

Falling within this scholarship, our book's main contribution lies in the effort to define and discuss some of the primary characteristics of the science/technology/market nexus as it manifests in Global South countries. The preceding sections have explained why and how we study technoscience and market expansion in the Global South. Through innovation, production, extraction, consumption, and revaluation practices, the Global South countries, institutions, and populations contribute to global technological capitalism. Simultaneously, technological expansion profoundly affects social, political, and biological life in the Global South. Furthermore, considering the entanglement of technologisation and capital expansion in the Global South sheds light on dynamics often overlooked when studying technoscientific capitalism solely from the perspective of the North. This section posits our analytical stance by outlining the major features that characterise technological globalisation in the Global South and differentiating it from developments in the Global North. Existing literature and our own research indicate several dynamics that characterise technoscientific globalisation in the Global South compared to high-income countries. The following paragraphs introduce eight main features, which are discussed in the subsequent chapters.

Firstly, technological globalisation in Global South countries is often characterised by the heterogeneity of equipment and infrastructures coexisting in the same space. Whether for water procurement (Anand 2011; Zérah 2014), electricity (Guillou and Girard 2023), or transport (Gopakumar 2020), a wide array of equipment coexists within the same territories, from brand new and cost-intensive materials to very old and low-tech ones. The same applies for technological devices. For instance, on Indian roads, one can encounter a great variety of means of transport, from expensive cars to cycle rickshaws, bullcarts, and camels. This heterogeneity and the hybridisation processes it fosters are partly addressed by the concept of 'creole technologies' proposed by historian David Edgerton (2007). Edgerton frames as creole technologies those that have been translocated, appropriated, and locally manufactured. One example Edgerton provides is that of long-tailed boats with car engines attached to wooden structures, turning them into speedboats, which first appeared in Bangkok and then spread to other Thai cities and Southeast Asian countries. However, this heterogeneity also results from the uneven distribution of resources among users and citizens, as explained by Ashley Carse while following Panamanian roads. While the canal hosts prosperous trade, for some people, 'infrastructure [be it roads or electricity] arrives slowly, if at all' and villages or whole areas can remain bluntly excluded from infrastructure installation (Carse 2014). Heterogeneity thus manifests through forms of prioritisation and feelings of exclusion. It relates to factors such as socioeconomic inequalities and weaker regulatory intervention by the state. It can also accompany decentralising dynamics, as in the case of energy procurement.

Secondly, the study of technological capitalism in the Global South high-lights alternative value-making strategies and specific market constructions that cater to demands absent in the richest countries. One example is 'Bottom of the Pyramid' market strategies (Prahalad 2005), which create forms of capital accumulation by involving the poorest segments of the world population in markets. In the financial sector this phenomenon is illustrated by financial inclusion, microfinance or microinsurance, along with the development of solar batteries, mini-grids, and low-cost electricity devices in the energy sector. (Guillou and Girard 2023; Sarkar 2021). The nutritional and medical sectors also see the emergence of specific goods, such as humanitarian fixes or nutra-ceuticals targeting the rural poor (Redfield 2016; Street 2015). Moreover, these market strategies align with various corporate formations and practices of capitalism that a focus solely on the richest countries would overlook. The role of some of the wealthiest businesspeople and corporations based in emerging countries must be thoroughly analysed in a world superficially dominated by Google, Apple, Facebook, Amazon, and Microsoft (the GAFAM). Reliance Industries in India, one of the largest petrochemical industries globally, is also a leading actor in biotechnologies and telecommunications. The Brazilian firm Vale S.A, one of the world's largest mining companies, plays a crucial role in renewing extractive practices and defining energy policies from Canada to New Caledonia. In Kenya, Safaricom has been pivotal in designing and implementing digital payment systems. The Vodacom Group, including Safaricom, and the Nedbank Group created M-PESA, aiming to provide a quick, safe, and easy way of transferring money via mobile phone technology, which was soon adopted by a large portion of the Kenyan population. These alternative markets and styles of corporate capitalism are supported by regulatory strategies that can differ from those promoted by the highest-intensive technological capitalist firms. Such differences are illustrated at times by disagreements and protests in arenas like the World Trade Organisation that reveal differing conceptions of regulation framed within Global South nations, particularly around intellectual property rights.

Thirdly, technoscientific globalisation progression in Global South countries accompanies the emergence of original and influential forms of governance. Due to their strategic positions as nodes and gatekeepers in international relations and economic exchanges (Cooper 2005), public administrations have evolved into hubs of industrial development and innovation (Mavhunga 2018). For this reason, governance in the South has been exposed to a greater extent, and much earlier than in the North, to computer science, statistics, and digital technologies (Zimmermann 1984; Breckenridge 2014). Recently, the digital industry has contributed to develop forms of governance made for the South, deeply affecting in return the development of governance and identification practices in both North and South. For example, multiple African countries were already equipped with biometric identity card systems, years before European countries developed biometric identity cards to identify their citizens. The digital industry has played a major role in the construction of such circulations from Southern to Northern governance. The Indian Aadhaar system is a particularly famous illustration of this model. Beyond Aadhaar, systems like Pakistan's National Database and Registration Authority, and others developed in Kenya, Ghana – and throughout Africa, especially West Africa, due to migration and Islamist attacks – demonstrate the widespread impact of such initiatives. Rooted in remote governance, these systems enable the provision and distribution of public services, but they are materialising very diverse politics, from financial inclusion to police surveillance or citizenship participation (Jacobsen 2012). These artifacts (the IT systems) fundamentally alter the relationship between governments and citizens, shifting towards 'coded citizenship', where individual characteristics are converted into data for administrative manageability (Kitchin 2016; Rao and Nair 2019; Taylor and Broeders 2015), thus digitalising the 'biopolitical technology of rule' (Masiero and Shakhti 2020; Rose 1999). While digital identity schemes may enhance food security governance, they can also introduce errors that exclude populations from essential services, thereby reducing access to citizenship (Eyenga et al. 2022; Nayak 2020), perpetuating North-South inequalities (Debos and Desgranges 2023), and displacing historical controversies over politics and public participation with technical debates centred on infrastructures and IT experts (Passanti and Pommerolle 2022).

Fourthly, technoscientific globalisation in Global South countries is marked by the heavy constraints imposed by international inequalities, as well as efforts to contest and circumvent these limitations. Individuals and institutions often depend on external funding, face restrictions on international mobility, and encounter asymmetries in decision-making power, largely due to significant differences in state funding and investment capacities. This dynamic is evident in global health research processes, where US and European institutions seek 'partnerships' in Global South countries but frequently make unilateral decisions on research topics, collaborators, and publication venues (Crane 2013; Feld and Kreimer 2019). Conversely, Global South actors must navigate constrained strategies to secure partnerships and opportunities (Kingori and Gerrets 2016). In response, these actors increasingly participate in new geopolitical assemblages and trade routes, distancing themselves from the economic and cultural dominance of the wealthiest countries. The COVID-19 pandemic provided striking examples of this reconfiguration of global trade relations and geopolitical formations. While European countries and the US monopolised most of the vaccine doses produced by their companies, Chinese and Russian COVID-19 vaccine developers established a significant presence in the developing world, employing business strategies that emphasised registration, collaborations, and manufacturing agreements, including full production processes in the South. Chinese COVID-19 vaccines accounted for nearly half the total volume produced globally in 2021[3]; nearly 900 million doses of Sinovac's CoronaVac were administered through vaccine cooperation agreements with 20 countries outside China, with the largest proportions going to Indonesia, Brazil, Turkey, and Chile. Most Sinovac agreements to expand manufacturing capacity were with upper-middle-income countries, with Egypt, Indonesia, and Brazil being the largest manufacturers by volume. Beyond COVID-19 vaccines, China is increasingly seen as a valuable partner by other Global South countries like Brazil, due to less stringent intellectual property requirements in trade and investment agreements compared to the US, EU, and Japan. This shift could facilitate new forms of agreements, leveraging negotiations with other global players (Ido 2023).

Fifthly, the detrimental social and environmental externalities of globalisation are largely borne by the Global South countries and populations. Air pollution and global warming effects are most severe in Global South countries (Khandekar et al. 2023; Landrigan et al. 2017), toxic materials such as dangerous pesticides or waste are more present due to the 'unequal distribution of exposure' (Bureau-Point 2021; Geissler and Prince 2020), and occupational health risks are higher in developing countries compared to the richest ones. Outsourced industrial activities often lead to outsourced pollutants. For instance, while many countries consume Indian medicines, this comes at the cost of significant environmental pollution due to chemical waste (Boudia et al. 2021). Among various case studies, the informal sector's role in collecting and salvaging waste, and reintegrating recovered materials into the formal economy, is noteworthy. Waste from Global North countries is seen as a resource in the Global South, where its collection is a vital economic activity and the recovery of resulting resources is key to the global economy. In many ways, some developing countries have become dumping grounds for technological waste from around the world, no longer limited to high-income countries (Lepawsky 2018). Among these numerous dumping grounds, Agbogbloshie's e-waste site in Ghana is one of the most infamous (Akese and Little 2018). Since Chinese authorities enforced stricter regulations, many used or obsolete items are increasingly sent to other parts of Asia or African countries for repair or dismantling, forcing vulnerable communities to live alongside highly polluting activities that constitute 'toxic colonialism' (Liboiron 2021; Pratt 2010). The academic field of waste studies has highlighted how domestic waste becomes an informal income source for many vulnerable people (mostly from lower socioeconomic classes) in the Global South countries, who are relied upon by institutional collection systems but struggle for political recognition (Bercegol and Gowda 2018; Dinler 2018; O' Hare 2019). This is illustrated by Indian waste pickers who reintegrate technological waste into value chains (Bercegol 2020). When Northern countries discard batteries and solar panels, complex and efficient valuation networks are established in countries like Kenya to repair and maintain existing solar grid technologies (Cross and Murray 2018).

Sixthly, technological development in the Global South, because devices and equipment still often come from the North, heavily involves technological adjustment and adaptation: modifying tools to meet local needs, adapting metrics to local standards, and tailoring food products to local tastes and nutritional requirements. Companies adjust their offer to local demands. For instance, Hindustan Lever Limited developed an affordable washing powder requiring less water (Subrahmanyan and Gomez-Arias 2008). However, customers themselves often initiate these processes by imaginatively repurposing technologies. This was evident with the Chinese domestic electrical manufacturer Haier, which redesigned washing machines to accommodate washing potatoes after realising that rural Chinese consumers were using their machines for this purpose in addition to doing laundry. Haier's engineers modified the machines by providing wider pipes to prevent clogging by mud (Anderson and Billou 2007). Technological innovations thus emerge not only through top-down, global-local, and unidirectional exchanges but as a back-and-forth process involving continuous adjustments. These situations, however, remain embedded in postcolonial legacies that often perpetuate power relations. Consequently, adjustment or adaptation can also take on conflictual forms, requiring stakeholders from the Global South to find arrangements – sometimes with global firms and sometimes against them – to overcome technological dependency. One of the most significant examples described in the literature is the process of reverse engineering or learning by copying, which involves extracting knowledge and know-how from a manufactured object. This has been studied in the context of generic antiretroviral (ARV) production in Brazil (Cassier and Correa 2009). ARV reverse engineering in Brazil has demonstrated the appropriation and adaptation of technology by a Global South country despite tight intellectual property rules and the global HIV epidemic.

Seventhly, this also makes us more aware of market features that, although they might exist worldwide, take on particular importance in resource-scarce settings in developing countries. This is the case of uncodified practices frequently unregulated by the state. The significance of informal maintenance and innovation practices is particularly high in countries where people have limited access to expensive standardised products. The Indian concept of 'jugaad' – which refers to a quick, improvised fix often seen in the informal sector, driven by urgent needs and limited resources (Kumar and Bhaduri 2014) – perfectly illustrates this (Bhaduri 2016; Parthasarathy 2022; Philip et al. 2012) as it shows specific practices of maintenance in developing contexts as compared to the richest countries (Denis and Pontille 2022). Another aspect of informality involves the complexity of informal sourcing processes within formal industrial operations, as highlighted by technological institutions and firms (Gameiro and Quet 2023). Informal sourcing is not exclusive to Global South countries, as it can also be found in the richest countries (Tsing 2015). However, it is a more systematic feature in developing contexts, where informal labour is often highly developed and where informal labourers are much more exposed to health and safety risks. This indicates that technological capitalism is intertwined with traditional hierarchies and forms of exclusion, such as the Indian caste system, which tend to be overlooked when focusing solely on the formal sections of value chains. Indian Dalit and tribal communities significantly contribute to global technological value chains, although much of their participation is invisibilised and poorly rewarded financially. In its more brutal form, technological capitalism feeds this informalisation through the mechanism of a 'depletion economy' (Precarity Lab 2020).

Lastly, the focus on technological developments in the Global South also calls for more attention to 'cosmotechnics', which refers to the ways in which technological developments resonate with situated conceptions and imaginaries of technological development (Hui 2021). Over centuries, a universalist framework was imposed by European nations through colonisation, sustained intellectually by the establishment of a 'colonial library' (Mudimbe 1988) and sometimes forcefully through various developmental or environmental interventions (Blanc 2024; Mavhunga 2018). This framework, however, tends to obscure the complex representations of the relationships between science, technology, and the universe, as well as their global diversity. Cosmotechnics, as a descriptive concept, highlights the existence of various understandings of the relations between science and society across continents, calling for a better appreciation of their role in national technoscientific projects, closely related to the notion of 'technoscientific imaginaries' (Jasanoff and Kim 2009). An increasing number of studies have focused on specific Southern countries as case studies. For instance, Anne Pollock immersed herself in a laboratory in Johannesburg, demonstrating how contemporary research and development practices in biotechnologies are anchored in a collective desire to see South Africa regain its status as an innovative power (Pollock 2019). Other studies have shown that 'traditional' medicine can be considered mainstream and modern in certain regions of Asia and Africa (Kloos 2020; Osseo-Asare 2014) and can become a field of major innovations in the pharmaceutical sector through processes of reformulation (Pordié and Gaudillière 2014). This last line of research invites focus on the effects of ethics, norms, values, and ideologies on the technoscientific evolutions of the Global South.

PRESENTATION OF CHAPTERS

Building on this analytical stance, the case studies gathered in this book shed light on the diverse ways in which social groups from the Global South engage in the expansion of technological markets by navigating complex power structures and regulatory frameworks, thereby contributing to shaping the trajectory of technoscientific capitalism. Methodologically, most of the work presented here is grounded in ethnographic approaches, and relies upon semi-structured interviews, direct and participatory observations, archival work, and engagement with grey literature. Geographically, the fieldwork has been anchored mainly in nine countries (Brazil, Chad, China, Ecuador, Ghana, India, Kenya, Mauritius, Senegal) on three continents – Africa, Asia, and Latin America – through ten chapters, which could be deemed either too few or too many: our focus is on attempting to define regularities in Global South experiences of technoscientific globalisation, and this book aims at being a first step in a longer research process. The geographical diversity of our studies is not primarily a strategic choice, and it essentially reflects our own individual fieldworks and area specialisations. The contingency of this selection accounts for the relative abundance of chapters studying India (3) while regions such as Southeast Asia or Oceania have been left aside. In this regard, our endeavour should be interpreted overall as a call for more dialogue and discussion between studies in multiple non-Western locations, aiming at the description of connections and similarities between Global South experiences of technoscientific globalisation. Such discussion could also put more emphasis than we have done – through lack of means – on the particular role of Chinese actors in these interconnections. Most of us were met during our fieldwork with the presence of Chinese actors but were not able to follow their leads since we were locally focused, saving this work for later. We think, however, that by discussing a variety of technologies and locations, our case studies help foster understanding of how technoscientific globalisation works from below – by taking seriously the features defined in this introduction.

Although all the chapters sporadically address several of the features exposed in the preceding section, we could not cover them systematically and had to make choices in order to explore in greater depth analytical points that were particularly important to understand our case studies. Following this, the chapters revolve around four main themes that contribute to defining the specificity of technoscientific globalisation in Global South countries.

The first theme this book explores is uncodified (by governments) practices and processes within the context of technoscientific globalisation. The authors delve into neglected aspects of digital capitalism to uncover the intricacy of informal practices within global technological value chains and technology-driven capitalist accumulation. Their arguments highlight the complexities of power relations and the ways in which technological capitalism is articulated within existing hierarchies and forms of exclusion.

Henry Chávez and María Belén Albornoz explore how migrants in Latin America divert the use of platform technologies, specifically focusing on the grey areas around (interchangeable) driver profiles and on the innovative ways in which migrants use them. Chávez and Belén Albornoz reveal how platform companies take advantage of informal and illegal labour workforce and social vulnerabilities to ensure the functionality of their technologies. By doing so, the authors shed light on the power dynamics and exploitative practices underlying the operationalisation of platform technologies – without neglecting to highlight the practices of resistance to such dynamics.

Javed Mohammad Alam, studying migrant workers in a South Asian context, explores how digital technology is adopted by Indian migrants in the Gulf countries, particularly to send money back home and maintain connections with their families in India. He highlights how the migrant condition and experience encourage the adoption of financial and digital technologies back in their home countries, showcasing the ways in which migrants engage with and use digital technologies for financial purposes. This brings him to insist upon the contribution of unskilled migrants to the global circulation of knowledge and expertise.

The second theme explored in this book is the power dynamics and constraints triggered by international inequalities within technoscientific globalisation. The authors delve into digital capitalists' and other global actors' participation in technology-driven capital accumulation, invisible and free labour, and data extraction. Their arguments highlight the complexities of power relations and the ways in which dominant actors shape technoscientific practices, especially in the digital and pharmaceutical sectors.

Jessica Pourraz and Allison Felix Hughes analyse how Ghanaian scientists have managed to build expertise in air pollution with imported low-cost sensors (LCS), which they adapt to the local setting and which thus compete with the highly technical and expensive measurement instruments from the Global North considered as the 'gold standard'. By exploring how Ghanaian scientists carry out LCS calibration to make the devices accurate and to ensure the validity of the data produced, the authors show how calibration work is, in turn, appropriated and valued on the market by LCS companies who sell their devices to other Global South countries. In doing so, the authors highlight the free and invisibilised contributions of Ghanaian actors to technological air monitoring infrastructure, showing the power dynamics between local actors and technology owners from the Global North.

Grounding their study in the context of the COVID-19 epidemics, Koichi Kameda, Denise Pimenta, and Gustavo Matta analyse how the vaccine was initially supplied to Brazil and document the public health authorities' subsequent journey towards achieving full production autonomy and vaccinal sovereignty. By analysing the benefits of technological transfers for the countries and actors from which the technology originates – such as Sinovac in China and AstraZeneca in the UK – they illustrate how technological transfers serve the interests of specific actors and result in the emergence of new actors in the Global South. The authors also highlight the role of local values and nationalistic ideologies in driving Brazil's efforts to achieve vaccine autonomy and cope with international inequities, exploring how these factors shape the development and use of technology.

The case study presented by Thibaut Serviant-Fine shows how China has taken up heparin production during a period of public health crisis and global procurement outsourcing, a process that has entailed a shift in the supply of farmed pig materials for the production of biological drugs. His chapter shows how the increasingly important role played by China has subsequently been put into question during zoonotic episodes, coupled with anxieties on the part of rich countries' governments about the growing power of Chinese firms in global markets. The chapter addresses the entangled role of procurement practices, animal product sourcing, and geopolitical relations in the fabrication of contemporary globalisation. By analysing the increasing and contested role of China, the author nuances the predominant image of South-to-North extractive circulations and supply circuits to demonstrate how technological and regulatory change at a global scale can also arise from actors not located in the Global North.

The third theme explored in the book is the local adaptation and adjustment practices surrounding globalised technoscientific tools. Drawing on the growing STS scholarship on non-Western countries, including the Global South, the authors examine how local actors creatively engage with and transform technologies to meet their specific local needs. By going beyond the dominant narratives of technological expansion from powerful sites in the Global North, the contributions highlight the active role of local actors in shaping technoscientific practices. The case studies explored by the respective authors shed light on different aspects of the local adaptation and innovation of digital and pharmaceutical technologies.

Marine Al Dahdah and Mathieu Quet's chapter documents the Mauritian logistics industry through the prism of its digital turn. The role of logistics is to manufacture 'seamless' commodity flows, and digital devices have been increasingly mobilised in order to perform this function. However, manufacturing seamlessness digitally requires continuous efforts of adjustment between global standards, hegemonic tools, and local resources. The chapter therefore examines the adaptation of digital software to meet local needs in Mauritius, emphasising the efforts of a local IT company to patch global platforms and local firms' environments in order to cater to the specific requirements of regional logistics. By doing so the chapter explores how digital technologies contribute to stabilising logistical flows at the cost of permanent patching work.

Studying pharmaceutical technologies in Chad, Ilyass Mahamat Nour Moussa shows how transnational actors such as international organisations play an essential role in the global integration of the Chadian pharmaceutical market by supporting the securitisation of medicine flows. In his case study, the author examines how the Chadian Ministry of Health seized the opportunity presented by a development programme funded by the World Bank, intended to reduce gender inequality in the Sahel countries, to set up a laboratory for controlling the quality of industrial medicine. By taking seriously the agency of the Chadian actors to bypass the norms implemented by international organisations, the author helps us to think about how local players are appropriating, repurposing, and circumventing standardised apparatuses to fit their specific needs – which reflect key dynamics of technoscientific globalisation 'from below'.

The final theme explored in the book relates to alternative value-making strategies and specific market fabrication within the context of technoscientific globalisation. The authors dig into the creation of markets in the Global South and investigate how in this context technology answers specific demands as much as it fashions expectations. By looking at the emergence of specific technological markets, the authors encompass a wide range of actors, corporate formations, strategies, and practices of capitalism that do not exist *per se* in the Global North. These different ways of shaping and forming markets are also supported by alternative regulatory strategies, values, and representations, which are also scrutinised by the authors.

Yves-Marie Rault-Chodankar shows how business practices in the Indian generic drugs industry contrast with the strategies of the IP-capital-rich industry in the Global North. He highlights companies' use of inventive copying to gain patient loyalty in low-income markets, to diversify their manufacturing capacities through national and international certifications to cater to diverse production needs, and to develop local distribution networks through multiple types of relationships with suppliers, wholesalers, retailers, and doctors. Rault-Chodankar considers whether the capitalist model proposed by Indian generic companies establishes a structured alternative to IP-based pharmaceutical globalisation.

Cecilia Passanti's contribution investigates the collaboration between public administrations and IT companies in developing technological tools to secure voting and democracy in Africa. Although biometrics are used worldwide for various purposes (migration control, passport issuance), they are particularly developed for and within Global South contexts in the case of elections. This is especially true in Africa, where many countries have formed long-term relationships with the IT industry to create, since independence, the material conditions for democracy, citizenship, and public participation. Passanti analyses the structured organisation of labour required to produce electoral biometric artefacts in Kenya and Senegal, which involves the daily work of institutions and individuals at both national and transnational levels. In doing so, she highlights how these technologies travel and become global through the reproduction, in many sites, of situated social relations of production.

Aamod Utpal investigates the processing of peanut butter to make a humanitarian good to address food security needs. The author examines how Global North actors, by redefining peanut butter as a medicine, have created a new market for ready-to-use therapeutic food (RUTF) producers, the majority of which are Indian. He explores the ways in which technological transformations influence market shaping and dynamics, and questions how the Indian localisation of RUTF participates in shaping specific answers and critiques to the product itself.

By addressing these themes, the book aims to uncover the agency of local actors, power dynamics, and the transformative potential of technology in shaping societies and markets in the Global South. By focusing on the experiences and innovations of actors from the Global South, we seek to challenge the ethics and implications of globalised industries, thereby encouraging critical discourse on alternative models of innovation, development and participation (Jasanoff 2004; Tyfield 2012), and contributing to a more inclusive and nuanced understanding of technoscientific globalisation.

THINKING TOGETHER, MAKING STS GLOBAL

This edited volume is the outcome of the activity of a group of young researchers united by a shared interest in analysing the production, circulation, and appropriation of scientific and technological objects in Africa, Asia, and Latin America. Coming together in 2019 at the initiative of Mathieu Quet (who was initially the only tenured scholar in the group) to think about science and technology in the Global South, our collective has been mostly composed of eleven scholars who were attached at one moment of their career to the Centre Population et Développement (CEPED) research unit in Paris, part of the broader Institut de Recherche pour le Développement (IRD).

More than half of our fellows were born and raised, and studied, in Global South countries (Chad, India, Brazil and Ecuador). The remaining five are natives of France and Italy, two of whom are the only women left of the group. At the time of publishing, there are only two permanent researchers, who turn out to be two of the three French white male scholars of the group. The rest of us are PhD candidates and/or post-doctoral research fellows searching for permanent positions and less precarious futures. Along this vibrant journey, two colleagues had to leave academia due to the lack of professional opportunities. One joined an NGO, and the other became a bartender: research can definitely lead to everything.

At the very beginning of our collective, we aimed to create a collaborative space for our heterogeneous group located in different classes, races, genders, levels of seniority, professional trajectories, and geographies (Harding 2004). We faced issues with bringing ourselves up to speed, especially when knowledge, language, theories, and epistemological frameworks mostly originate from Europe and the US. The difficulties experienced by our two Indian members in navigating French academia and visa administration without any prior knowledge in French illustrated the complexities of a supposedly postcolonial world in which it remains extremely hard for Southerners to get visas and housing in the face of systemic racism and due to massive differences in living cost between their countries of origin and Europe. The arrival among us of a Chadian PhD candidate, fresh off the plane in France and with no English skills, further illuminated some of the major issues entangled in epistemic inequities and injustices, not only because English dominates the academic world but also because there had been no opportunity in Chad to study STS at master's level. It also made communication within the group more complicated and forced us to imagine and implement alternative translation practices – with which we fondly experimented.

We met regularly between 2019 and 2024 to discuss the inner workings of what we call 'technoscientific globalisation from below'. Applying critical reflexivity to our methodological practices allowed us to address the imperial legacy of extractive research practices (Liboiron 2021: 1). Through our research, we embraced decolonising approaches by aiming to bring to light historically silenced voices, to acknowledge and present the people we are writing about as knowledge holders and, beyond this, knowledge makers (Thambinathan and Kinsella 2021). By being attentive to the power relations between us, as colleagues, and between us as researchers and the people we met during our fieldwork, we critically reflected on our own practices, sharing the responsibility to open up space for decolonising visions (Thambinathan and Kinsella 2021). This was made possible by centring the concerns and worldviews of Global South individuals when studying the globalisation of technological goods in poorer, non-hegemonic contexts (Smith and Hanson 2012). Basing our method upon the collection of empirical work, we aimed at giving a voice to the plethora of actors from developing and emerging countries whose active role in the production, circulation, regulation, appropriation, and consumption of technologies is critical to current globalisation processes – even though it is frequently dismissed and silenced. By doing so, we intended to curb the perpetuation of inaccurate stereotypes about Global South countries regarding technological globalisation.

From 2021, we started to organise, once a year, short-duration stays, from three days to one week, during which we experienced more fully an academic life organised around collaborative work, writing, and living. We worked hard during the day to craft the chapters of this book while it was still a project, but as a group we also self-organised in informal and pleasant ways: deconstructing hierarchies and tasting French wine, Brazilian caipirinhas, or Indian Old Monk rum during lively apéritifs, cooking together and sharing food, discussing the merits of Bollywood movies, inquiring about Chadian politics, and often ending up dancing to the tune of whoever had appropriated the Bluetooth speaker.

We also decided to open the conversation to a broader pool of researchers and set up a monthly online seminar entitled 'Technological globalisation from below,' with the support of the CEPED, the Global Research Institute of Paris (GRIP), and the Université Paris Cité. From 2021 to 2024 this seminar became an opportunity to discover, welcome, and listen to the work of researchers in the humanities and social sciences, half of them originating from Global South countries, half of them women, whose research topics were in the Global South and who anchored their studies in the field of science and technology studies (STS). We were trying as much as we could to fight against the imbalance between 'mainstream' and 'peripheral' STS (Invernizzi et al. 2022). We wish to warmly thank these researchers for the great value of their contributions, which deeply informed the reflections presented in this book: Festus Boamah (Dept of Geography, Universität Bayreuth), Xan Chacko (Brown University), Armelle Choplin (Dépt de géographie et environnement, Université de Genève), Kevin P. Donovan (Centre of African Studies, University of Edinburgh), Abena Dove Osseo-Asare (Dept of History, University of Texas), Stefan Ecks (School of Social and Political Science, University of Edinburgh), Christine Ithurbide (CNRS, Passages), Nathalie Jas (INRAE, IRISSO), Sibel Kusimba (Dept of Anthropology, University of South Florida), Joshua Lepawsky (Dept of Geography, Memorial University), Nancy Odendaal (University of Cape Town), Alyssa Paredes (Dept of Anthropology, University of Michigan), Anne Pollock (Global Health & Social Medicine, King's College), Laurent Pordié (CNRS, Cermès3), Rajeswari Raina (International Relations and Governance Studies, Shiv Nadar University), Luisa Reis-Castro (Dept of Anthropology, University of Southern California), Suvobrata Sarkar (Rabindra Bharati University, Kolkata), Puleng Segalo (University of South Africa), James H. Smith (Anthropology Dept, UC Davis), and Gabriela Soto Laveaga (History of Science Dept, Harvard University).

These discussions also permitted us to keep meeting online, as most of us were often doing fieldwork abroad and not regularly in touch with our common host university. More generally the seminar was a place to share and discuss our concerns regarding the major problem centred on epistemic inequities and injustices. On one occasion we notably received Professor Puleng Segalo from the University of South Africa to talk about their collaborative project to radically transform research partnerships between Africa and the Global North. Their Africa Charter on Transformative Research Collaborations (Gebremariam et al. 2023) is instrumental in repositioning Africa's global science and research ecosystem. It has become an inspirational tool for our collective to question why African ways of being and other knowledge systems in the Global South are often relegated to the periphery, and Euro-Western ways of knowing are centralised as the norm. We want to express our particular thanks to Puleng Segalo for her intervention.

At last, we decided that all these reflections deserved to be published together – and quite soon it appeared we would work with Mattering Press. We wish to thank the publishing team wholeheartedly for their help and support, with a particular shoutout to Anna, who followed the project with much friendliness and dedication. We also wish to thank Tabita Rezaire for accepting that her work would be used on the cover of the book. This means a lot to us as her work has questioned in various ways the coproduction of technology, colonialism, and race. We are very proud to slide our thoughts, literally, in between her visual reflections at the intersection of Afrofuturism and cyberfeminism that more generally irrigate our discussions. Nonta Libbrecht-Carey has played a key role in translating or editing most of the initial manuscript, which has mostly been written by non-native English speakers. With this book, we seek to shed light on the multitude of social actors occupying subaltern positions and contributing through multiple initiatives to innovate, to produce, but also to consume; we hope this intervention indicates how their participation increasingly gives shape to different forms of globalisation. Unearthing these untold stories may hopefully contribute to generating new imaginaries through which to invent alternative futures that are fairer and based on collective needs. The chapters unveil the possibilities, in an increasingly capitalistic world where inequities and hierarchies are still considered inevitable for globalisation, for everyone to have a role. Hence, we dedicate this book to all the actors we met in our respective research sites and who are endeavouring to make the Global

South something other than just a place of extraction, i.e., where innovations and/as alternatives are also conceived.

ENDNOTES

1 The expression 'Global South' – referring mostly to former colonies and poorer countries – marks a divide with the 'North', mostly formed of the richest nations and former Western colonial powers. Although the expression offers a binary and simplistic vision of the world, it tends to indicate structural power asymmetries and helps us throughout the text to take into account the critical consequences of such inequalities. We therefore assume the relevance of 'Global South' and praise the possibility of contestation it brings to the global order (Prashad 2012) in contrast to country nominalism or other categorisations such as the Goldman Sachs–coined 'BRICS' or the World Bank's 'emerging markets', which not only include fewer countries but also focus primarily on perspectives of economic growth.

2 Joseph Needham pictured the history of science as multiple civilizational rivers running into the same ocean of knowledge – science.

3 Data available via the Knowledge Portal on innovation and access to medicines: https://www.knowledgeportalia.org/covid-19 (accessed June 2024).

REFERENCES

Abraham, I., *The Making of the Indian Atomic Bomb: Science, Secrecy and the Postcolonial State* (Zed Books, 1998).

Aga, A., *Genetically Modified Democracy: Transgenic Crops in Contemporary India* (Yale University Press, 2021).

Akese, G.A. and Little, P.C., Electronic waste and the environmental justice challenge in Agbogbloshie. *Environmental Justice*, 11 (2018) pp.77–83 https://doi.org/10.1089/env.2017.0039.

Anderson, J., and N. Billou, 'Serving the World's Poor: Innovation at the Base of the Economic Pyramid', *Journal of Business Strategy*, 28.2 (2007): 14–21.

Anderson, W., and V. Adams, 'Pramoedya's Chickens: Postcolonial Studies of Technoscience', in U. Felt, R. Fouché, and others, eds., *The Handbook of Science and Technology Studies*, 3rd edn (MIT Press, 2008), pp. 181–204.

Anand, N., Pressure: the politechnics of water supply in Mumbai. *Cultural Anthropology: Journal of the Society for Cultural Anthropology*, 26(4) (2011), pp.542–564. https://doi.org/10.1111/j.1548-1360.2011.01111.x.

Appadurai, A., *Modernity at Large: Cultural Dimensions of Globalization* (University of Minnesota Press, 1996).

Arnold, D., *Everyday Technology: Machines and the Making of India's Modernity* (University of Chicago Press, 2019).

Bercegol, R. de, and S. Gowda, 'A New Waste and Energy Nexus? Rethinking the Modernisation of Waste Services in Delhi', *Urban Studies* (2018) https://doi.org/10.1177/0042098018770592.

Bhaduri, S., *Frugal innovation by the 'small and the marginal': An alternative discourse on innovation and development.* Inaugural lecture: Erasmus University Rotterdam. (2016) Available at: https://www.eur.nl/en/news/frugal-innovation-small-and-marginal-pcc-inaugural-lecture

Blanc, G., *La nature des hommes: Une mission écologique pour « sauver » l'Afrique* (La Découverte, 2024) https://hal.science/hal-04504506/.

Boamah, F., 'Desirable or Debatable? Putting Africa's Decentralised Solar Energy Futures in Context', *Energy Research & Social Science*, 62 (2020): 101390 https://doi.org/10.1016/j.erss.2019.101390.

Boczkowski, P. J., *Abundance: On the Experience of Living in a World of Information Plenty* (Oxford University Press, 2021).

Boudia, S., and others, *Residues: Thinking Through Chemical Environments* (Rutgers University Press, 2021).

Bouquillion, P., C. Ithurbide, and T. Mattelart, eds., *Digital Platforms and the Global South: Reconfiguring Power Relations in the Cultural Industries* (Routledge, 2023) https://doi.org/10.4324/9781003391746.

Breckenridge, K., *Biometric State* (Cambridge University Press, 2014).

Bureau-Point, È., *Les Patients Experts dans la Lutte Contre le Sida au Cambodge: Anthropologie d'une Norme Globalisée* (Presses universitaires de Provence, 2021).

Buu-Sao, D., Participating as indigenous people. Government and protest in the Amazonian uses of consultation – Peru. *Participations*, 2019/3(25), pp.33–58. (2019) https://shs.cairn.info/journal-participations-2019-3-page-33?lang=en

Carse, A., *Beyond the Big Ditch: Politics, Ecology, and Infrastructure at the Panama Canal* (MIT Press, 2014).

Cerutti, S., 'Who is Below?', *Annales. Histoire, Sciences Sociales*, 70.4 (2015): 931–55 https://doi.org/10.1353/ahs.2015.0167.

Cassier, M. and Correa, M., eds., *Health innovation and social justice in Brazil.* (Palgrave Macmillan, 2018). https://doi.org/10.1007/978-3-319-76834-2.

Chacko, X., *Invisible Vitality: The Hidden Labours of Seed Banking*, in J. Bangham, X. Chacko, and J. Kaplan, eds., *Invisible Labour in Modern Science* (Rowman & Littlefield Publishers, 2022), pp. 217–25.

Chee, L. P., *Mao's Bestiary: Medicinal Animals and Modern China* (Duke University Press, 2021).

Choplin, A., *Concrete City: Material Flows and Urbanization in West Africa* (John Wiley & Sons, 2023).

Choplin, A., and O. Pliez, 'The Inconspicuous Spaces of Globalization', *Articulo – Journal of Urban Research*, 12 (2015) https://doi.org/10.4000/articulo.2905.

Choplin, A., and O. Pliez, *La Mondialisation des Pauvres: Loin de Wall Street et de Davos* (Média Diffusion, 2018).

Cooper, F., *From Colonial State to Gatekeeper State in Africa* (Mario Einaudi Center for International Studies, 2005) https://hdl.handle.net/1813/55008.

Crane, J. T., *Scrambling for Africa: AIDS, Expertise, and the Rise of American Global Health Science* (Cornell University Press, 2013).

Cross, J., and D. Murray, 'The Afterlives of Solar Power: Waste and Repair Off the Grid in Kenya', *Energy Research & Social Science*, 44 (2018): 100–109 https://doi.org/10.1016/j.erss.2018.04.034.

de Bercegol, R., 'From Trash to Cash, Recovering Practices, Wholesale Markets and Industrial Recycling in Delhi', *Science Journal of Architecture & Construction – Tạp Chí Kiến Trúc & Xây Dựng*, 38 (2020): 21–27.

De la Cadena, M., *Earth Beings: Ecologies of Practice Across Andean Worlds* (Duke University Press, 2015).

De Laet, M., and A. Mol, 'The Zimbabwe Bush Pump: Mechanics of a Fluid Technology', *Social Studies of Science*, 30.2 (2000): 225–63.

Debos, M., and G. Desgranges, 'L'invention d'un marché: Économie politique de la biométrie électorale en Afrique', *Critique Internationale*, 98.1 (2023): 117–39 https://doi.org/10.3917/crii.098.0117.

Denis, J., and D. Pontille, *Le Soin des Choses: Politiques de la Maintenance* (La Découverte, 2022).

Dinler, D. Ş., 'New Forms of Wage Labour and Struggle in the Informal Sector: The Case of Waste Pickers in Turkey', in J. Pattenden, L. Campling, and others, eds., *Class Dynamics of Development* (Routledge, 2018), pp. 90–110.

Donko, K., *Territory, Power and Politics at a Frontier in Central Benin* (Nomos, 2022).

Donovan, K. P., and E. Park, 'Knowledge/Seizure: Debt and Data in Kenya's Zero Balance Economy', *Antipode*, 54.4 (2022): 1063–85 https://doi.org/10.1111/anti.12815.

Dumoulin Kervran, D., M. Kleiche-Dray, and M. Quet, 'Going South. How STS Could Think Science in and with the South?', *Tapuya: Latin American Science, Technology and Society*, 1.1 (2018): 280–305 https://doi.org/10.1080/25729861.2018.1550186.

Edgerton, D., *The Shock of the Old: Technology and Global History Since 1900* (Oxford University Press, 2007).

Edgerton, D., *Britain's War Machine: Weapons, Resources, and Experts in the Second World War* (Oxford University Press, 2011).

Eyenga, G. M., G. O. Omgba, G. O, and J. F. Bindzi, 'Être Sans-Papier Chez Soi? Les Mésaventures de l'Encartement Biométrique au Cameroun', *Critique Internationale*, 97.4 (2022): 113–34.

Feld, A., and P. Kreimer, 'Scientific Co-Operation and Centre-Periphery Relations: Attitudes and Interests of European and Latin American Scientists', *Tapuya: Latin American Science, Technology and Society*, 2.1 (2019): 149–75.

Gago, V., *Neoliberalism from Below: Popular Pragmatics and Baroque Economies* (Duke University Press, 2017).

Gameiro, M. B. P., and M. Quet, 'Feral Pharmaceuticalization—Biomedical Uses of Animal Life in Light of the Global Donkey Hide Trade', *BioSocieties*, 18.3 (2023): 679–706 https://doi.org/10.1057/s41292-022-00288-2.

Gebremariam, E. B., I. A. Aboderin, D. Fuh, and P. Segalo, 'Beyond Tinkering: Changing Africa's Position in the Global Knowledge Production Ecosystem', *CODESRIA Bulletin*, 7 (2023).

Geissler, P. W., and R. J. Prince, 'Layers of Epidemy: Present Pasts during the First Weeks of COVID-19 in Western Kenya', *Centaurus*, 62.2 (2020): 248–56 https://doi.org/10.1111/1600-0498.12295.

Gopakumar, G., *Installing Automobility: Emerging Politics of Mobility and Streets in Indian Cities* (MIT Press, 2020).

Gudynas, E., *Extractivisms: Politics, Economy and Ecology* (Fernwood Publishing, 2020).

Guillou, E., and B. Girard, 'Mini-Grids at the Interface: The Deployment of Mini-Grids in Urbanizing Localities of the Global South', *Journal of Urban Technology*, 30.2 (2023): 151–70 https://doi.org/10.1080/10630732.2022.2087170.

Hailwood, M., and B. Waddell, eds., 'The Future of History from Below: An Online Symposium' (2013).

Harding, S. G., *The Feminist Standpoint Theory Reader: Intellectual and Political Controversies* (Psychology Press, 2004).

Harding, S., *Sciences from Below: Feminisms, Postcolonialities, and Modernities* (Duke University Press, 2008) https://doi.org/10.2307/j.ctv11smmtn.

Harding, S., ed., *The Postcolonial Science and Technology Studies Reader* (Duke University Press, 2011).

Hecht, G., *Being Nuclear: Africans and the Global Uranium Trade* (MIT Press, 2014).

Hui, Y., *Art and Cosmotechnics* (University of Minnesota Press, 2021).

Ido, V. H. P., 'From "Threat" to "Partners"? The Changing Landscape of Innovation and Intellectual Property between China and Brazil', in M. H. Ueta, M. Alencastro, and R. Pinheiro-Machado, eds., *How China Is Transforming Brazil* (Springer Nature, 2023), pp. 59–78 https://doi.org/10.1007/978-981-99-3102-6_4.

Invernizzi, N., G. Foladori, and D. Maclurcan, 'Nanotechnology's Controversial Role for the South', *Science, Technology and Society*, 13.1 (2008): 123–48.

Invernizzi, N., A. Davyt, L. R. Medina, and P. Kreimer, 'STS between Centers and Peripheries: How Transnational Are Leading STS Journals?', *Engaging Science, Technology, and Society*, 8.3 (2022): 31–62.

Jacobsen, K. L., 'Unique Identification: Inclusion and Surveillance in the Indian Biometric Assemblage', *Security Dialogue*, 43.5 (2012): 457–74 https://doi.org/10.1177/0967010612458336.

Jasanoff, S., ed., *States of Knowledge: The Co-Production of Science and Social Order* (Routledge, 2004).

Jasanoff, S., and S.-H. Kim, 'Containing the Atom: Sociotechnical Imaginaries and Nuclear Power in the United States and South Korea', *Minerva*, 47.2 (2009): 119–46.

Kaplinsky, R., J. Chataway, N. Clark, and others, 'Below the Radar: What Does Innovation in Emerging Economies Have to Offer Other Low-Income Economies?', *International Journal of Technology Management & Sustainable Development*, 8.3 (2009): 177–97 https://doi.org/10.1386/ijtm.8.3.177/1.

Khandekar, A., J. Cross, and A. Maringanti, 'Scale and Modularity in Thermal Governance: The Replication of India's Heat Action Plans', *Urban Studies*, 0.0 (2023).

Kingori, P., and R. Gerrets, 'Morals, Morale and Motivations in Data Fabrication: Medical Research Fieldworkers' Views and Practices in Two Sub-Saharan African Contexts', *Social Science & Medicine*, 166 (2016): 150–59 https://doi.org/10.1016/j.socscimed.2016.08.019.

Kitchin, R., *Getting Smarter about Smart Cities: Improving Data Privacy and Data Security* (Data Protection Unit, Department of the Taoiseach, 2016), p. 82.

Kloos, S., 'Humanitarianism from Below: Sowa Rigpa, the Traditional Pharmaceutical Industry, and Global Health', *Medical Anthropology*, 39.2 (2020): 167–81.

Kumar, H. and Bhaduri, S., 'Jugaad to grassroots innovations: Understanding the landscape of the informal sector innovations in India'. *African Journal of Science, Technology, Innovation and Development*, 6(1) (2014) pp.13–22. https://doi.org/10.1080/20421338.2014.895481.

Kusimba, S., *Reimagining Money: Kenya in the Digital Finance Revolution* (Stanford University Press, 2021).

Landrigan, P. J., and others, 'The Lancet Commission on Pollution and Health', *The Lancet*, 391.10119 (2017): 462–512 https://doi.org/10.1016/S0140-6736(17)32345-0.

Langwick, S. A., *Bodies, Politics, and African Healing: The Matter of Maladies in Tanzania* (Indiana University Press, 2011).

Laveaga, G. S., 'Largo Dislocare: Connecting Microhistories to Remap and Recenter Histories of Science', *History and Technology*, 34.1 (2018): 21–30.

Lei, Y.-W., *The Gilded Cage: Technology, Development, and State Capitalism in China* (Princeton University Press, 2023).

Lepawsky, J., *Reassembling Rubbish: Worlding Electronic Waste* (MIT Press, 2018).

Liboiron, M., *Pollution Is Colonialism* (Duke University Press, 2021).

Lindtner, S. M., *Prototype Nation: China and the Contested Promise of Innovation* (Princeton University Press, 2020).

Lobato, R., *Netflix Nations: The Geography of Digital Distribution* (New York University Press, 2019) https://doi.org/10.18574/nyu/9781479895120.001.0001.

Losego, P., and R. Arvanitis, 'La science dans les pays non hégémoniques', *Revue d'anthropologie des Connaissances*, 2.3 (2008) https://doi.org/10.3917/rac.005.0334.

Martinez-Alier, J., *The Environmentalism of the Poor: A Study of Ecological Conflicts and Valuation* (Edward Elgar Publishing, 2003).

Masiero, S., and S. Shakhti, eds., 'Unique Identification in India: Aadhaar, Biometrics and Technology-Mediated Identities', *South Asia Multidisciplinary Academic Journal*, 23 (2020).

Mathews, G., G. L. Ribeiro, and C. A. Vega, *Globalization from Below: The World's Other Economy* (Routledge, 2012).

Mavhunga, C. C., *The Mobile Workshop: The Tsetse Fly and African Knowledge Production* (MIT Press, 2018).

Mbembe, A., *Brutalisme* (La Découverte, 2020).

Medina, E., I. da Costa Marques, and C. Holmes, *Beyond Imported Magic* (MIT Press, 2014) https://mitpress.mit.edu/9780262526203/beyond-imported-magic/.

Mudimbe, V. Y., *The Invention of Africa: Gnosis, Philosophy, and the Order of Knowledge* (Indiana University Press, 1988).

Nayak, N., 'Chasing Rights in Delhi: Social Movements and the National Food Security Act', *South Asia Multidisciplinary Academic Journal*, 23 (2020) https://doi.org/10.4000/samaj.6306.

Needham, J., 'Science and Society in East and West', *Science & Society*, 28.4 (1964): 385–408.

O'Hare, P., '"The landfill has always borne fruit": Precarity, Formalisation and Dispossession among Uruguay's Waste Pickers', *Dialectical Anthropology*, 43.1 (2019): 31–44 https://doi.org/10.1007/s10624-018-9533-6.

Odendaal, N., *Disrupted Urbanism: Situated Smart Initiatives in African Cities* (Bristol University Press, 2023).

Osseo-Asare, A. D., *Bitter Roots: The Search for Healing Plants in Africa* (University of Chicago Press, 2014).

Osseo-Asare, A. D., *Atomic Junction* (Cambridge University Press, 2019).

Paredes, A., 'Experimental Science for the "Bananapocalypse": Counter Politics in the Plantationocene', *Ethnos*, 88.4 (2023): 837–63 https://doi.org/10.1080/00141844.2021.1919172.

Parthasarathy, S., 'How Sanitary Pads Came to Save the World: Knowing Inclusive Innovation through Science and the Marketplace', *Social Studies of Science* 52.5 (2022) https://doi.org/10.1177/03063127221122457.

Passanti, C., and M.-E. Pommerolle, 'The (Un)Making of Electoral Transparency through Technology: The 2017 Kenyan Presidential Controversy', *Social Studies of Science*, 52.6 (2022): 928–53.

Peschard, K., *Seed Activism: Patent Politics and Litigation in the Global South* (MIT Press, 2022).

Peterson, K., *Speculative Markets: Drug Circuits and Derivative Life in Nigeria* (Duke University Press, 2014).

Philip, K., L. Irani, and P. Dourish, 'Postcolonial Computing: A Tactical Survey', *Science, Technology, & Human Values*, 37.1 (2012): 3–29 https://doi.org/10.1177/0162243910389594.

Pollock, A., *Synthesizing Hope: Matter, Knowledge, and Place in South African Drug Discovery* (University of Chicago Press, 2019) https://doi.org/10.7208/9780226629216.

Pordié, L., and J.-P. Gaudillière, 'The Reformulation Regime in Drug Discovery: Revisiting Polyherbals and Property Rights in the Ayurvedic Industry', *East Asian Science, Technology and Society: An International Journal*, 8.1 (2014): 57–79.

Powell, J., *The Survival of the Fitter: Lives of Some African Engineers* (Intermediate Technology Publications, 1995) https://cir.nii.ac.jp/crid/1130282271254351872.

Prahalad, C. K., *The Fortune at the Bottom of the Pyramid* (Wharton School Publishing, 2005).

Prasad, A., *Science Studies Meets Colonialism* (John Wiley & Sons, 2022).

Prashad, V., *The Poorer Nations: A Possible History of the Global South* (Verso, 2012).

Pratt, L. A. W., 'Decreasing Dirty Dumping – A Reevaluation of Toxic Waste Colonialism and the Global Management of Transboundary Hazardous Waste', *Texas Environmental Law Journal*, 41 (2010): 147.

Precarity Lab, *Technoprecarious* (Goldsmiths Press, 2020).

Quet, M., *Illicit Medicines in the Global South: Public Health Access and Pharmaceutical Regulation* (Routledge, 2022).

Rajan, K. S., *Pharmocracy: Value, Politics, and Knowledge in Global Biomedicine* (Duke University Press, 2017).

Rao, U., and V. Nair, 'Aadhaar: Governing with Biometrics', *South Asia: Journal of South Asian Studies*, 42.3 (2019): 469–81 https://doi.org/10.1080/008564 01.2019.1595343.

Redfield, P., 'Fluid Technologies: The Bush Pump, the LifeStraw® and Microworlds of Humanitarian Design', *Social Studies of Science*, 46.2 (2016): 159–83 https://doi.org/10.1177/0306312715620061.

Rose, N., *Powers of Freedom: Reframing Political Thought* (Cambridge University Press, 1999) https://doi.org/10.1017/CBO9780511488856.

Rottenburg, R., *Far-Fetched Facts: A Parable of Development Aid* (MIT Press, 2009).

Sabbagh, Dan, 'Houthis Call West's Bluff with Renewed Red Sea Drone Assault', *The Guardian*, 10 January 2024.

Said, E., 'Orientalism Reconsidered', in B. J. Moore-Gilbert, ed., *Postcolonial Criticism* (Routledge, 1997), pp. 126–44.

Sarkar, S., *Let There Be Light: Engineering, Entrepreneurship and Electricity in Colonial Bengal, 1880–1945* (Cambridge University Press, 2021).

Smith, J. H., *The Eyes of the World: Mining the Digital Age in the Eastern DR Congo* (University of Chicago Press, 2021).

Smith, R. D., and K. Hanson, eds., *Health Systems in Low- and Middle-Income Countries: An Economic and Policy Perspective* (Oxford University Press, 2012).

Stephenson, N., 'Emerging Infectious Disease/Emerging Forms of Biological Sovereignty', *Science, Technology, & Human Values*, 36.5 (2011): 616–37 https://doi.org/10.1177/0162243910388023.

Street, A., 'Food as Pharma: Marketing Nutraceuticals to India's Rural Poor', *Critical Public Health*, 25.3 (2015): 361–72 https://doi.org/10.1080/09581596.201 4.966652.

Subrahmanyan, S., and J. Tomas Gomez-Arias, 'Integrated Approach to Understanding Consumer Behavior at Bottom of Pyramid', *Journal of Consumer Marketing*, 25.7 (2008): 402–12 https://doi.org/10.1108/07363760810915617.

Subramaniam, B., *Holy Science: The Biopolitics of Hindu Nationalism* (University of Washington Press, 2019).

Tarrius, A., *La Mondialisation par le Bas: Les Nouveaux Nomades des Économies Souterraines* (Balland, 2002).

Tastevin, Y. P., 'Autorickshaw (1948–2…): A Success Story', *Techniques & Culture*, 58 (2012): 264–77 https://doi.org/10.4000/tc.6306.

Taylor, L., and D. Broeders, 'In the Name of Development: Power, Profit and the Datafication of the Global South', *Geoforum*, 64 (2015): 229–37 https://doi.org/10.1016/j.geoforum.2015.07.002.

Thambinathan, V., and E. A. Kinsella, 'Decolonizing Methodologies in Qualitative Research: Creating Spaces for Transformative Praxis', *International Journal of Qualitative Methods*, 20 (2021): 16094069211014766 https://doi.org/10.1177/16094069211014766.

Thompson, E. P., *The Making of the English Working Class* (Vintage, 1966).

Tsing, A. L., *Friction: An Ethnography of Global Connection* (Princeton University Press, 2005) https://doi.org/10.1515/9781400830596.

Tsing, A. L., *The Mushroom at the End of the World: On the Possibility of Life in Capitalist Ruins* (Princeton University Press, 2015) https://doi.org/10.1515/9781400873548.

Tyfield, D., *The Economics of Science: A Critical Realist Overview*, volume 2, *Towards a Synthesis of Political Economy and Science and Technology Studies* (Routledge, 2012).

Waast, R., *20th Century Sciences* (6 volumes) (IRD Editions, 1995).

Waisbich, L. T., S. Roychoudhury, and S. Haug, 'Beyond the Single Story: "Global South" Polyphonies', *Third World Quarterly*, 42.9 (2021): 2086–95.

Williams, L. D. A., *Eradicating Blindness: Global Health Innovation from South Asia* (Springer Singapore, 2019) https://doi.org/10.1007/978-981-13-1625-8.

Zérah, M.-H., 'Politics and Governance in the Water Sector: The Case of Mumbai', in G. Schneier-Madanes, ed., *Globalized Water: A Question of Governance* (Springer Netherlands, 2014), pp. 197–210 https://doi.org/10.1007/978-94-007-7323-3_14.

Zimmermann, J.-B., 'Politiques Africaines de l'Informatique', *Politique Africaine*, 13.1 (1984): 79–90 https://doi.org/10.3406/polaf.1984.3688.

LIVING OFF THE INFORMAL

2

ENABLING AND RESISTING THE PLATFORM ECONOMY FROM BELOW: PLATFORM IMMIGRANT WORKERS IN ECUADOR

Henry Chávez and María Belén Albornoz

FIG. 2.1 Platform workers looking at their smartphones for orders to deliver at Quito in Ecuador, 2021 (Henry Chávez and María Belén Albornoz)

'IT'S LIKE BEING IN A TOXIC RELATIONSHIP: SOMETIMES "SHE" TREATS YOU well; sometimes "she" punishes you and gives you nothing. You never know when, how or why', Elmer, a middle-aged man from Venezuela, explains as he points to his phone with the delivery app ['she'] for which he works.

He and three other colleagues have agreed to talk to us about their working conditions. They have been waiting fruitlessly for hours for an order to deliver, at the car park of a well-known fast-food restaurant in a high-class neighbourhood of Quito. It is almost noon, and the sun is beating down at 2,800 metres above sea level. All four are part of the half million immigrants who have arrived in Ecuador in the last six years, the largest migratory flow in the country's history. Only Elmer has formal immigration status – for the others, this is the only job they could get without a work visa.

Over the last decade, online platforms for gig work have become a new vector for the technological globalisation process. These platforms have transformed work around the world by removing several barriers, facilitating workers' access to the labour market, and providing a quick source of income. However, they have also brought harmful social and environmental externalities by undermining labour rights, reducing occupational health, increasing the use of disposable products, and creating new forms of exploitation, especially for vulnerable populations such as immigrants. This is the case in Ecuador, where the main delivery and ride-hailing platforms began operating around 2016 against a backdrop of economic recession resulting from falling oil prices and exports (the country's main source of income) and a significant influx of migrants. Many of these immigrants have difficulty finding formal employment, not only because of adverse economic conditions but also because of the constraints caused by international inequalities, such as the limitations on the international circulation of people embodied in administrative barriers to obtaining work visas, a lack of local contacts, and discrimination. Thus, the platform economy has become an alternative for some of these workers, offering them flexibility and income opportunities but at the cost of poor working conditions and new forms of exploitation.

Elmer's words somehow sum up a widespread sense of puzzlement among platform workers regarding the way in which the platforms allocate their orders, rates, compensations, and penalties. This idea of a 'toxic relationship' between the worker and the platform echoes the metaphor of the *pharmakon* (Stiegler 2007), an ancient Greek word used to designate at once the remedy, the poison, and the scapegoat of a disease. In the correct dose, this substance (or technology) could cure a person, but in excess, it becomes toxic and could kill them. Platforms seem to produce a similar effect. They help workers – many in critical situations (unemployed, without an income, immigrant) – to obtain an immediate and flexible source of income and somehow alleviate their condition. At the same time, these platforms put workers in a situation of dependence and vulnerability that forces them to accept poor working conditions and exposes them to new forms of disciplining and automated exploitation that are incomprehensible to the vast majority. The black box they hold in their hands thus appears to them as a capricious being that manages their time and rewards them as it pleases. They have no choice but to obey or cheat. In the words of José – another platform worker – 'here you don't have a boss; your boss is the platform'.

Platform companies rely heavily on algorithms to manage worker recruitment, task assignment, pricing, compensation, management, and evaluation. Although algorithms promise to improve efficiency and reduce costs, they also produce new forms of inequality, discrimination, exploitation, and injustice, as they can perpetuate bias and exclusion, oversimplify tasks and situations, or simply fail. A growing body of literature on platform economics and algorithmic management has already addressed some of these issues. This chapter attempts to build on these two approaches but focuses on a subaltern and vulnerable population: migrant platform workers in the Global South. In order to do that, we conducted about 115 semi-structured and 10 in-depth interviews, as well as ethnographic observations with platform workers in Ecuador between 2021 and 2023. Specifically, we sought to answer this question: How do workers learn to use and navigate these platforms, and how do they develop innovative tactics to circumvent their limitations or take advantage of their shortcomings?

The results of this research shed light on the specificities of the deployment of the platform economy and its algorithmic management systems in the Global South and how this process unveils some of the features of technological globalisation, as seen below. Specifically, they show how the interactions between workers and platforms through informal non-codified spaces and strategies allow them to resist and circumvent algorithmic management systems while enabling and expanding platforms' functioning (Gago 2017).

This chapter is organised as follows: The next section briefly overviews the platform studies and algorithmic management literature. We then present the Ecuadorian context and our methodology. Then, we explain the role of the black markets for platform profiles and vehicles, as well as workers' strategies of resistance to algorithmic management. This section is followed by a general discussion of subaltern human agency. Finally, we conclude by showing how vulnerable and subaltern populations, such as migrant workers, simultaneously enable and resist the globalisation of technological platforms from below.

PLATFORM STUDIES AND ALGORITHMIC MANAGEMENT FROM BELOW

The concept of the 'platform' has emerged in the last two decades following the irruption and success of companies such as Google, Facebook, Amazon, Uber, Tinder, and many others (Courtois and Timmermans 2018; van Dijck 2013). While the term was initially associated with video games (Montfort and Bogost 2009), it later spread to other digital media and infrastructure (Plantin et al. 2018). Today, it encompasses social media platforms (Langlois and Elmer 2013), streaming services (Lobato 2019), and online marketplaces (Ipeirotis 2010). Several case studies have shed light on the transformations brought about by these new complex sociotechnical objects, particularly Facebook (Bucher 2012; Ellison et al. 2007; Tufekci 2015), YouTube (Abidin 2018; Baertl 2018; Burgess and Green 2009), and Amazon (Oestreicher-Singer and Sundararajan 2012). More recently, other studies grouped under the subfield of the 'gig' or 'platform' economy have focused on ride-hailing platforms (de Freitas et al. 2023; Rosenblat and Stark 2016), delivery platforms (Galiere 2020; Lord et al. 2023; Timko and van Melik 2021), and crowd-work platforms (Casilli and Bouquin 2020; Irani 2015; Le Ludec et al. 2020; Tubaro et al. 2020). Several scholars from various disciplines have highlighted the potential benefits of platform work (flexibility, autonomy for workers, and reduced transaction costs for consumers), but also the risks (employment precariousness, lack of benefits and social protection, harmful externalities and informality) (Albrieu 2021; Woodcock and Graham 2019). Others have focused on the experiences from below of subaltern, marginalised, and vulnerable workers, such as immigrants. The latter face the constraints imposed by international inequalities, language barriers, a lack of social networks, and discrimination (van Doorn and Vijay 2021) but also promote collective action and the organisation of workers (Albornoz and Chavez 2020; Woodcock and Cant 2022) or alternative value-creation strategies, such as platform cooperatives, geared towards creating worker-owned and democratic platforms that prioritise worker wellbeing and community building (Grohmann 2021).

Scholars in other branches have investigated the algorithmic management systems that power those platforms (Firmino et al. 2019; Rosenblat and Stark 2016; Vallejos Rivero 2021). These systems are used to match workers with tasks, set prices, and evaluate performance. So-called algorithmic work (Jarrahi et al. 2021; Schildt 2017), designed to enable collaboration between algorithms and humans, can be observed within the framework of ride-sharing services, delivery services, and the execution of micro-tasks. However, algorithms can also create new forms of control and surveillance (Wood et al. 2019). Algorithmic management is often opaque, partly due to companies' secrecy regarding their systems but also because of the large amounts of data at stake, the complexity of the calculations, and the interactions between different algorithms involved in decision-making that render them very difficult for humans to fully understand (Faraj et al. 2018; Meijerink and Bondarouk 2023; Parent-Rocheleau and Parker 2022). Furthermore, workers may experience difficulty, confusion, and uncertainty in understanding whether they should comply with the algorithm's instructions and the potential consequences of non-compliance (Zarsky 2016). They may also perceive the algorithm as unfair, inaccurate, or restrictive, yet they may hesitate to override it (Markus 2017). These conditions can lead to resistance to algorithmic work (Kellogg et al. 2020).

Building on this literature, we unpack the experiences of immigrant workers in the platform economy in Ecuador. By examining how algorithmic management affects their labour experiences and the strategies they develop, we seek to understand how this subaltern and marginalised population enables such platforms to adjust, adapt, and operate while simultaneously, from below, circumventing and resisting exploitation and the process of disciplining. This sheds light on how non-hegemonic actors in the Global South use tactics unforeseen by the platforms' designers in the Global North to attempt to redefine the asymmetrical power relationship between platforms and workers, but, at the same time, they contribute to the expansion of the very same platform economy.

RECESSION, MIGRATION, AND THE PLATFORM ECONOMY IN ECUADOR

Ecuador is one of the smallest economies in South America and is structurally dependent on exports of primary goods, particularly oil. The lack of development of the non-extractive economy has resulted in high rates of poverty, inequality, unemployment, and informal employment. Since 2015, the country has faced an economic recession triggered by falling oil prices, rising public debt, and trade deficits (Burchardt et al. 2016). By the end of 2022, its GDP had barely returned to its 2015 level (Banco Central del Ecuador 2023). This recession has led to the growth of unemployment and the informal economy. In 2014, only 49% of the economically active population (about 8 million people) had an adequate job.[1] The COVID-19 pandemic and generalised lockdowns exacerbated this situation. By June 2020, about 84% of the population was unemployed, underemployed, or working in informal activities. The situation has improved since, but not enough to return to pre-pandemic levels: by February 2023, about 67% of the population was still unemployed, underemployed, or working in informal activities (INEC 2023).

Over the past ten years, Ecuador has experienced a significant influx of migrants, with more than 630,000 people arriving there. The majority of these migrants originate from Venezuela (58%), followed by Colombia (22%), Peru (8%), and Cuba (7%). However, the outbreak of the COVID-19 pandemic in 2020 marked a turning point, with approximately 100,000 migrants leaving the country up to 2021 (INEC 2022).

In this challenging context, platforms like Cabify, Uber, UberEats, Glovo, and Rappi started operating in Ecuador, providing a lifeline for those who had lost their jobs, or immigrants seeking work, by offering a relatively simple and quick way to earn a regular income. Despite the lack of infrastructure, limited connectivity, and relatively weak demand for online services (DataReportal 2023), these companies have decided to expand their operations to countries such as Ecuador. The absence of regulation and the growing pool of workers accustomed to working in low-paying jobs without legal benefits provide an opening for these companies to earn greater profits.

No official records are available to estimate the number of individuals working on platform economy apps or the proportion of migrants participating in such activities. However, based on various reports, public declarations, and interviews, we estimate that platform workers in Ecuador account for about 1% of the under-employed population, that is, around 40,000 workers, about 35% of whom are immigrants (Albornoz et al. 2022; CITEC 2022; Maya et al. 2022; OIT 2022).

Between 2021 and 2023, we carried out ethnographic observations and 115 semi-structured interviews, including 10 in-depth interviews with workers from seven platforms: Rappi, Glovo, PedidosYa (formerly Glovo), Uber Eats, Cabify, Uber, and Didi. Workers were contacted through the platforms or at known worker meeting points using a snowball strategy. We interviewed plat-form workers who could share their real-life experiences of using algorithmic systems. In this way we gained an in-depth understanding of their practices and the contextual circumstances of their work.

The ages of the interviewees ranged from 19 to 71, but more than two-thirds were less than 38 years old. Eight out of ten were men, four out of ten were immigrants, and more than half had a technical or university diploma. Cabify, Rappi, and Uber were the most used platforms (two out of ten), followed by Didi, PedidosYa, Uber Eats, and Glovo (one out of ten). For the 10 in-depth interviews, we chose the most willing respondents to contribute to the research: six immigrants and four locals.

The interviews aimed to uncover workers' latent understanding of algo-rithms as tools and managerial structures by focusing on their experiences with the platforms. The interview protocol included questions about the workers' overall experience with the platforms; how they use them; their understanding of the underlying algorithms; their perception of the job assignment, pricing, and evaluation systems; how they look for information about the platforms' functioning; and how they deal with algorithmic control.

In the following sections, we present the outcomes of the platforms' expansion into the Global South and how intermediaries and informal markets enable the emergence of alternative value-creation strategies, novel forms of exploitation, and other harmful social and environmental externalities. Our analysis emphasises the workers' capacity to interact with and resignify, adjust, and adapt to the platforms' technological frameworks and demands, leading to new forms of resistance and technological appropriation from below.

ENABLING THE PLATFORM ECONOMY FROM BELOW: INFORMAL MARKETS FOR PLATFORM PROFILES AND VEHICLES

Until 2021, Ecuador had about half a million immigrants (INEC 2022). A survey of about 6,800 immigrants in Quito suggests that only 20% of them have formal immigration status (a work visa or residence permit) (Célleri 2020). This explains why only 55% claim to be employed and less than half have a formal contract. Consequently, these vulnerable workers are frequently subjected to exploitation and labour rights violations: 50% of them report working more than the legal 40-hour work week, only 12% receive more than the minimum legal wage of US$400 per month, and 44% experience discrimination based on their nationality.

In these conditions, many migrants turn to platforms as a relatively easy and quick way to earn money. According to CITEC (2022), about 37% of platform workers are immigrants. Platforms differ from traditional employment providers by operating in a non-codified informal space. From the period of their emergence, since their organisation and business models were entirely novel and most of their operations virtual, they were completely unregulated. For example, platforms could accept foreign documents to create worker accounts, allowing migrants to work under their profiles. But as they grew, local governments started pushing for regulation. Although several of these regulations in Ecuador are still under discussion, the platforms have already restricted access to new workers who lack documentation (work visa, identity card, or driver's license). This strategy was to smooth out the new workers' still informal and even illegal presence in the market. Indeed, during the first few years of operations, the platforms offered their drivers legal advice to recover the vehicles and cover the costs of the fines. However, few drivers used this service because it was easier to bribe the police and continue working than to lose several days of work without the car.

As a result, the subcontracting or renting of platform profiles on the black market became a common practice among undocumented immigrant workers. Most of the interviewees in this study stated that they had used or were currently using rented accounts due to a lack of suitable documentation. This alternative strategy has allowed immigrants to circumvent administrative and algorithmic restrictions to access the job opportunities offered by the platforms, and for the latter to continue expanding their markets and extracting profits from the grey areas created by the lack of local regulation and uncodified processes. In other words, by redefining the asymmetrical relationship with platforms (from which they were initially excluded by design), through an alternative strategy unforeseen by the platforms' designers in the Global North, workers allow the platforms to have a larger workforce willing to accept their conditions and thus expand their operations. However, this strategy has also resulted in harmful externalities: a new system of exploitation that shifts risk from the employers to the employees, leading to unfair and unstable working conditions. Furthermore, many of these workers lack access to their vehicles, causing them to resort to a black market of 'investors', a pool of car or motorbike owners interested in renting out their vehicles or hiring drivers. Workers communicate directly with them to negotiate the terms and conditions of their agreements and subcontracting practices. This means that workers have to pay for both the vehicle and another person's platform profile.

According to our interviews, two types of arrangements exist: a fixed weekly fee or 50% of the earnings. The fixed weekly fee for a ride-hailing app ranges from US$75 to US$100. Our respondents reported working an average of 11 hours per day and earning approximately US$200 per week, much more than the legal 8-hour working day, which shows the degradation of occupational health. After paying for the car rental, they were left with only 50% of their income, which amounts to around US$400 per month, the minimum legal wage in Ecuador. The delivery workers experienced a similar situation. Those renting profiles are charged between US$15 and US$40 per week. If a delivery worker rents a profile and a motorbike, they may pay up to US$50 per week. Interviewees reported earning an average of US$150 per week or US$400 per month after expenses. Unlike with a regular job, this does not include any form of health or risk insurance, social security benefits, sick leave, vacation, or other job-specific expenses (helmet, smartphone, internet use, etc.)

The employment status of platform workers is a contentious issue, as they face numerous challenges, including a lack of labour rights, unstable income, a lack of benefits, and precarious and exploitative working conditions. Platforms claim that workers are self-employed entrepreneurs who manage their own time and have no employer. Informal subcontracting practices thus make it even harder for workers to secure minimum working conditions and social benefits. Those who rent their profiles act as intermediaries between undocumented immigrant workers and the platforms, enabling the latter to operate and expand. However, this exposes these workers to high risks and harmful externalities and shifts the cost of operating in a non-codified informal grey area, and of the platforms' expansion, onto their shoulders.

RESISTANCE FROM BELOW TO ALGORITHMIC MANAGEMENT: DEALING WITH A TOXIC RELATIONSHIP

Elmer's intuition about the 'toxicity' of the relationship with the platforms suggests a pharmacological problem (Stiegler 2007): a critical situation in which their survival depends on a technology that may be toxic to them. Here, we draw on our respondents' accounts and understanding of their situation to shed light on how subaltern actors resist, circumvent, and hack this technology to reduce or master its toxicity.

The term 'La toxica', which some of our interviewees (90% of them men) used to describe the app on their phones, reflects the embedded algorithms that govern these platforms and, consequently, the daily lives of platform workers. These algorithms, opaque and closed (Rosenblat and Stark 2016), serve as the black boxes managing the platforms. They assign orders and determine times, places, routes, prices, and payments. They also allocate rewards and penalties to ensure the supply of services and increase efficiency and quality.

> During the high-demand period, they ask you to deliver at least two or three orders … but if you do just one, they lower your score. I mean, it is not your fault but that of the restaurant […] They also punish you based on the customers. […] A bad score […] reduces 3 points your general score. (Interview, May 2021)

The implementation of these global platforms at the local level is intricate and frequently results in conflicts, prompting the apps to regularly update, adjust, and adapt to local conditions. This creates an unpredictable relationship between the platforms and the workers, who must strive to decipher and navigate the constantly evolving regulations and demands to maximise their earnings and avoid sanctions.

> December was crazy. One day, you were 'diamond'; later on, you were 'red'. They take you and give you points for no reason. […] The app has 508 updates so far. […] algorithm works very badly […] It sends you a triple order […] How can you be in three places at the same time? (Interview, February 2021)

The result is a kind of liquid Taylorism (Altenried 2020; Bauman and Lyon 2012) in which every worker has a planning office and a control system in their pocket but no workplace and no boss – at least not a human one.

> That's the problem. They don't care. They don't consider the time you lose when they make changes or updates without even warning you. Not an email, nothing. […] They just tell you, 'keep your score' […]. Everything is through messages. You have to upload a picture of the evidence of what is happening, and this is not an immediate procedure. There is no person to help you. (Interview, March 2021)

International inequality adds a layer of complexity to the analysis of algorithmic management presented above. Algorithmic scoring systems are designed to assign rewards or penalties to registered users who meet specific criteria, such as being legally eligible to work in a given country. Nevertheless, undocumented immigrants have found a way to enter the system through a black market of profiles and accounts, challenging the very foundations and assumptions of the scoring system. This Pavlovian algorithm aims to allocate, optimise, and evaluate tasks and payments to incentivise individual users and improve overall performance. However, rewards and penalties may not be distributed equally among profile owners and subcontracted workers.

> I've worked with three different accounts. [...] The owner of the first one told me, 'My friend, this is an excellent account. It will drop you six orders per day [...] That's why I ask you for US\$40 per week.' [...] I was fooled like a teenager in love. Then, another friend offered me a 'green' account [...], but this time, I said, 'Look, if I don't get any orders during the day, I'll give you back the account, and we keep being friends' [...] So, then, I asked a colleague, 'My friend, how does it work this thing? Can you show me?' [...] He explained that you have to 'upgrade' the account to 'diamond' in order to get more orders. [...] I tried for three more days, wasting fuel, going here and there, and I got nothing. So, I gave back this account [...] Then, I took another one. He told me, 'I will rent you the account and motorbike for US\$50.' It had three orders and 54 points. It is a very low number, as I know now, but at the time, I thought it was not so bad. I mean, it is far from zero, isn't it? [they laugh] (Interview, February 2021).

The profile black market used by undocumented immigrants has resulted in the incentives of the algorithmic management system being split across and shared by multiple individuals with vastly different backgrounds and behaviours. The rewards are shared between profile owners and migrant workers, who end up being subject to a disciplinary algorithmic punishment regime but for a lower reward than regular users. Thus, while the penalties and physical effort are borne solely by the worker, the profile owners reap only the rewards. Ultimately, undocumented workers bear the brunt of the problems associated with this system, as their daily lives are controlled and influenced by the Pavlovian mechanism.

> You must try to do things correctly, so the algorithm helps you improve. [...] The application has some things you have to respect. You have to do this and that and do things correctly, such as deliver on time, be polite, and talk to the customer. (Interview, March 2021)

However, workers can also adjust and adapt to algorithms by altering and observing their data collection processes to better utilise some parts of the system. By grasping the inner workings of certain platform features, workers can enhance the performance of certain platform features or repurpose them altogether.

> There are things I didn't know. Because I was new, I started to understand by searching the internet. I found that the application likes you to talk to the customer, even if the customer doesn't answer. [...] The algorithm keeps [your score] stable, but if you start by being rushed, if you don't write to the customer, the score starts to get lower. (Interview, March 2021)

By looking at the experience of undocumented immigrants, we can see that the effects of algorithmic management can go beyond the platforms themselves. In a kind of Black Mirrorian twist, the machine has broken free of its black box to exercise disciplinary measures not only on those who are officially registered in its database but also on those who should not be part of it. Here again, we can see how alternative strategies of non-hegemonic actors in the Global South attempting to change an asymmetrical power relationship end up contributing to the expansion of the platform economy beyond its limits in an unforeseen way. These insights show us some features and deviations of the technological globalisation process that can only be seen from below, which a Global North–based perspective would have missed.

SUBALTERN HUMAN AGENCY VS. PLATFORMS' TECHNOLOGICAL POWER

Despite the opacity of algorithmic management technologies, it is important to highlight that they still rely heavily on non-automated human work, as several scholars have noted (for example, Casilli et al. 2019). Moreover, the dynamics linked to migration shed light on the offline human labour that underpins the platform economy. By examining these dynamics at the frontiers of the global system, we can better understand how subaltern populations appropriate, reinterpret and modify the new technological regime imposed on them by the platforms, revealing not only the importance of human labour in the digital age but also the different paths that technological globalisation may take as it moves into the Global South.

Although getting started on platforms may seem simple, they are intricate systems with numerous automated services. Workers invest a significant amount of time and energy in comprehending the platform's ratings, deadlines, policies, and procedures (Mohlmann and Zalmanson 2017). To make sense of the complex workings of platforms, they might turn to social channels like WhatsApp chat groups to share knowledge and experiences with others. These chat groups serve as a resource for workers to seek advice on dealing with demanding customers, understanding companies' policies, and staying up to date with platform changes. Some workers also learn directly from more experienced colleagues.

Sensemaking can benefit workers by providing them with a better understanding of algorithms and algorithmic management, which increases their confidence and sense of control. However, algorithms may also limit workers' ability to perform their tasks. In such cases, platform workers may look for ways to avoid algorithmic processes or substitute them with outside tools. As they discover glitches in the algorithms, they may also learn to use these glitches to their advantage.

App-based platforms in peripheral contexts such as Ecuador rely on social practices and interactions beyond their algorithms. For example, the prevalence of immigrant workers on online platforms can be partially attributed to the convergence of the platforms' need for cheap and flexible labour with the immigrants' desire for accessible income, as they can work as much as they want. However, before this mutually beneficial relationship could be established, platforms had to overcome two significant obstacles. First, immigrant workers had to have access to the means of production: vehicles and phones. Second, platforms had to operate in a non-codified, informal, grey, unregulated, or even illegal space, which posed significant risks. In order to overcome these barriers, the platforms had to switch to an analogue approach and do extensive offline work and lobbying to ensure the viability of their algorithms. This was the case of the first ride-hailing platform that entered the Ecuadorian market. Like traditional analogue companies, this platform had to set up a physical office to meet potential candidates, provide them with training, and give promotional talks. It engaged in lobbying and reached informal agreements with workers, investors, and authorities, which were crucial to their start-up processes. These activities were eventually reduced after early adopters were secured, but the platform still supports a profile black market that operates outside the official system. This market creates a loophole in the algorithmic management system, resulting in lower earnings for immigrant workers at the end of the chain.

Far from being passive subjects, however, immigrants learn to deal with the platforms' demands and innovate around them. Driven by their needs and hopes, they work over ten hours daily, follow the platform's rules, try to comply with its algorithmic management, and even pay for its mistakes. But they also try to understand the platform, learn from it, and when possible, outsmart it. An interesting example of this was a sort of social hacking that a group of drivers carried out on one of the ride-hailing platforms operating in Quito. The route between the city and the airport was known to be one of the most profitable rides. However, the platform's algorithm prevented individual drivers from doing only that route. This limitation was reinforced by the fact that drivers were only offered a ride if they were close to the pick-up point, so if they went to the airport, they would not be able to get another ride immediately from there unless they waited, which was not possible due to the traffic regulations in place at the airport and because the platform was officially illegal. This group of drivers, therefore, joined forces to create a kind of cooperative group that shared several smartphones with multiple profiles and parked at a gas station near the airport to get rides from there to the city. This social hack allowed them to get a large share of the rides between the city and the airport and to multiply their incomes beyond what the platform would foresee.

> We started a sort of group, and we mounted a 'platform' down at the air-
> port. We made US$2,000 per month. However, this was because we did

something [the platform] could never imagine [...] because the app lets you have another phone. [My second phone] was in a queue at the airport. So, when I arrived there, I already had another customer waiting. That was very profitable. (Interview, February 2021)

By harnessing technology and using it to their advantage, workers, especially immigrants, can resist the control imposed by algorithmic management and try to redefine an asymmetrical power relationship. This innovative approach is driven by the same force that compelled these people to leave their home country and seek a better life in a new place. However, the very same strategy and actors contribute to the expansion of the platform market, ensuring a constant flow of drivers willing to take customers to and from the airport at their own risk. It should not be forgotten that the use of platforms is still illegal in Ecuador and that police controls at airports are frequent.

FIG. 2.2 Platform drivers detained by the police near Quito airport during social protests in October at Quito, Ecuador, 2019 (Ministry of the Interior of Ecuador)

This was, in fact, the cause that put an end to such a subaltern enterprise. At the height of the 2019 social revolt, the police arrested 19 platform drivers near the airport (mostly immigrants) and falsely accused them of planning an attack on the president. Later, it was revealed that they were only platform drivers parked 'unusually' near the airport (El Universo 2019; Vásconez 2020). This event put the platforms under the media spotlight and raised questions about how these still 'illegal' platforms were being used, especially around airports, where formal cab companies have an exclusive right to work. Shortly after, the platform updated its algorithm and removed the possibility of having the same profile on two different phones, closing the possibility of using the system invented by this group of drivers. Once again, this demonstrates how the strategies of subaltern actors to rebalance asymmetric relations with techno-transnational power also have an impact on the way technology is designed, adjusted, and globalised, even if it is against their interests and will. This particular feature highlights the importance of considering these subaltern populations' role in the local adoption, appropriation, and resignification of global technologies.

CONCLUSIONS

The platform economy is one of the main vectors of the global dissemination of digital technologies and the attending datafication and algorithmic management process. It has removed several traditional barriers to employment, facilitated workers' access to the labour market, and provided a quick source of income. However, these transformations have also undermined labour conditions and created new forms of exploitation and other harmful externalities, especially for vulnerable populations such as immigrants. Based on the case of immigrant platform workers in Ecuador, we have explored how subaltern actors deal with this process of technological globalisation from below.

Drawing on the growing literature on platforms, platform economics, and algorithmic management, and based on an ethnographic approach and more than 100 interviews with platform workers conducted between 2021 and 2023, we have shown how these workers interpret, interact, adapt, and resist the technological framework and demands of platforms and how the modern globalisation process driven by these platforms forces them to adjust, adapt, and negotiate their practices and technology in complex and unforeseen contexts. Platforms do so by taking advantage of the lack of regulation and the uncodified informal grey areas and practices in which they operate. These informal practices and areas that platforms use and help to perpetuate are the cradle of the detrimental externalities that the globalisation process brings to the Global South and one of the critical elements to understand the process of globalisation from below.

The main finding of this research is that in the context of international inequality, subaltern populations such as immigrant platform workers at once resist and enable the process of technological globalisation brought about by the platform economy. Indeed, workers are not merely passive recipients of algorithmic management and control systems imposed by those platforms. They develop strategies to learn, circumvent, adjust, and adapt to the algorithmic management exercised by the platform. By engaging in sensemaking activities, users acquire a working understanding of algorithms and their potential effects on their work. Through these sensemaking strategies, workers gain enough familiarity with the platforms' functions to effectively work with and around them, even opening up the black box of algorithms used by digital platforms. However, by developing alternative strategies, such as informal black markets for platform profiles and vehicles, to circumvent legal, administrative, and algorithmic barriers, they enable platforms to keep working and expanding their operations from below. By contributing to the development of these non-codified (informal) grey areas, they have given new breath to the development of these platforms in the South. Despite the different conditions under which they operate in the Global North, for which they were initially conceived, it is very likely that marginalised actors in this context also engage in similar practices, but this will require further enquiry.

Finally, workers may also use their learning to circumvent the algorithms or manipulate them to their advantage. They can bypass algorithmic constraints by drawing on other resources and technologies or by using their knowledge of algorithms to achieve desired outcomes. However, platform technologies are not static and tend to adapt, adjust, and incorporate changes to regain power over workers' practices. Consequently, a platform's algorithms, which are essentially an automated representation of the platform organiser's interests, are not deterministic rules. They iterate and are redesigned through feedback loops. In that sense, workers appropriate and expand the digital platform's system of programmatic processes as part of an information infrastructure. Platform workers and users have agency and influence over the platform's algorithms, its design, and redesign iterations. Algorithmic management is a sociotechnical process that results from continuous interactions between algorithms and humans. Workers' encounters with algorithms shape not only their work practices but also algorithmic outcomes and decision-making, which fuel the globalisation of platform technologies.

ENDNOTES

1 Working 40 hours a week for a minimum wage of US$400 per month, having a formal contract, and being registered with the social security system. Underemployment and informal employment include all jobs that do not meet these conditions.

REFERENCES

Abidin, C., *Internet Celebrity: Understanding Fame Online* (Emerald Publishing, 2018).

Albornoz, M. B., and H. Chavez, 'De La Gestión Algorítmica Del Trabajo a La Huelga 4.0', *Mundos Plurales – Revista Latinoamericana de Políticas y Acción Pública*, 7.2 (2020): 43–54 https://doi.org/10.17141/mundosplurales.2.2020.4848.

Albornoz, M. B., H. Chavez, and others, *Fairwork Ecuador Ratings 2022: Labour Standards in the Platform Economy* (Fairwork, 2022) https://fair.work/en/fw/publications/labour-standards-in-the-platform-economy-ecuador-ratings-2022/.

Albrieu, R., ed., *Cracking the Future of Work: Automation and Labour Platforms in the Global South* (FOWIGS, CIPPEC, IDRC, 2021) https://fowigs.net/publication/cracking-the-future-of-work-automation-and-labour-platforms-in-the-global-south/.

Altenried, M., 'The Platform as Factory: Crowdwork and the Hidden Labour Behind Artificial Intelligence', *Capital and Class*, 44.2 (2020): 145–58 https://doi.org/10.1177/0309816819899410.

Baertl, M., 'YouTube Channels, Uploads and Views: A Statistical Analysis of the Past 10 Years', *Convergence: The International Journal of Research into New Media Technologies*, 24.1 (2018): 16–32 https://doi.org/10.1177/1354856517736979.

Bauman, Z., and D. Lyon, *Liquid Surveillance: A Conversation* (Polity Press, 2012).

Banco Central del Ecuador, *Informe Estadístico Mensual* (October 2023).

Bucher, T., 'Want to Be on the Top? Algorithmic Power and the Threat of Invisibility on Facebook', *New Media & Society*, 14.7 (2012): 1164–80 https://doi.org/10.1177/1461444812440159.

Burchardt, H.-J., R. Domínguez, C. Larrea, and S. Peters, *Nada Dura Para Siempre: Neo-Extractivismo Tras El Boom de Las Materias Primas* (UASB – ICDD, 2016).

Burgess, J., and J. Green, 'The Entrepreneurial Vlogger: Participatory Culture Beyond the Professional/Amateur Divide', in P. Snickars and P. Vonderau, eds., *The YouTube Reader* (National Library of Sweden, 2009), pp. 89–107.

Casilli, A., and S. Bouquin, 'Il N'y a Pas D'automatisation Sans Micro-Travail Humain', *Les Mondes Du Travail*, 24–25 (2020): 3–21.

Casilli, A., P. Tubaro, and others, *Le Micro-Travail En France: Derrière L'automatisation, de Nouvelles Précarités Au Travail? [Rapport Final]*, Projet DiPLab 'Digital Platform Labour' (2019) http://diplab.eu/.

Célleri, D., *Situación Laboral y Aporte Económico de Inmigrantes En El Centro/Sur de Quito-Ecuador*(Rosa Luxemburg Stiftung, 2020) http://www.rosalux.org.ec/pdfs/SituacionLaboralYAporteEconomicoDeInmigrantes.pdf.

CITEC, *Estudio de Impacto de Las Plataformas Digitales En Ecuador* (Cámara de Innovación y Tecnología Ecuatoriana, 2022).

Courtois, C., and E. Timmermans, 'Cracking the Tinder Code: An Experience Sampling Approach to the Dynamics and Impact of Platform Governing Algorithms', *Journal of Computer-Mediated Communication*, 23.1 (2018): 1–16 https://doi.org/10.1093/jcmc/zmx001.

DataReportal, *Digital 2023: Ecuador* (2023) https://datareportal.com/reports/digital-2023-ecuador.

de Freitas, C., P. A. G. Pascoal, A. da S. Mello, and L. F. Silva, 'Digital Platform: Uber', *Revista Tecnologia E Sociedade*, 19.55 (2023): 176–88 https://doi.org/10.3895/rts.v19n55.14645.

El Universo, 'Extranjeros Detenidos En Aeropuerto de Quito Tenían Agenda Presidencial, Según Ministra de Gobierno', *El Universo* (10 October 2019) https://www.eluniverso.com/noticias/2019/10/10/nota/7554917/extranjeros-detenidos-aeropuerto-quito.

Ellison, N. B., C. Steinfield, and C. Lampe, 'The Benefits of Facebook "Friends": Social Capital and College Students' Use of Online Social Network Sites', *Journal of Computer-Mediated Communication*, 12 (2007): 1143–66 https://doi.org/10.1111/j.1083-6101.2007.00367.x.

Faraj, S., S. Pachidi, and K. Sayegh, 'Working and Organizing in the Age of the Learning

Algorithm', *Information and Organization*, 28.1 (2018): 62–70 https://doi.org/10.1016/j.infoandorg.2018.02.005.

Firmino, R. J., B. de V. Cardoso, and R. Evangelista, 'Hyperconnectivity and (Im) Mobility: Uber and Surveillance Capitalism by the Global South', *Surveillance & Society*, 17.1–2 (2019): 205–12 https://doi.org/10.24908/ss.v17i1/2.12915.

Gago, V., *Neoliberalism from Below: Popular Pragmatics and Baroque Economies* (Duke University Press, 2017).

Galiere, S., 'When Food-Delivery Platform Workers Consent to Algorithmic Management: A Foucauldian Perspective', *New Technology, Work and Employment*, 35.3 (2020): 357–70 https://doi.org/10.1111/ntwe.12177.

Grohmann, R., 'Rider Platforms? Building Worker-Owned Experiences in Spain, France, and Brazil', *South Atlantic Quarterly*, 120.4 (2021): 839–52 https://doi.org/10.1215/00382876-9443392.

ILO, *El Trabajo En Las Plataformas Digitales de Reparto y Transporte En Ecuador* (International Labour Organization, 2022).

INEC (Instituto Nacional de Estadística y Censos), 'Entradas y Salidas Internacionales', *Instituto Nacional de Estadística y Censos* (31 May 2022) https://www.ecuadorencifras.gob.ec/entradas-y-salidas-internacionales-informacion-historica/.

INEC (Instituto Nacional de Estadística y Censos), 'Estadísticas Laborales', *Instituto Nacional de Estadística y Censos* (February 2023) https://www.ecuadorencifras.gob.ec/estadisticas-laborales-abril-2022/.

Ipeirotis, P. G., 'Analyzing the Amazon Mechanical Turk Marketplace', *XRDS: Crossroads, The ACM Magazine for Students*, 17.2 (2010): 16–21 https://doi.org/10.1145/1869086.1869094.

Irani, L., 'Difference and Dependence among Digital Workers: The Case of Amazon Mechanical Turk', *South Atlantic Quarterly*, 114.1 (2015): 225–34 https://doi.org/10.1215/00382876-2831665.

Jarrahi, M. H., G. Newlands, and others, 'Algorithmic Management in a Work Context', *Big Data & Society*, 8.2 (2021): 20539517211020332 https://doi.org/10.1177/20539517211020332.

Kellogg, K. C., M. A. Valentine, and A. Christin, 'Algorithms at Work: The New Contested Terrain of Control', *Academy of Management Annals*, 14.1 (2020): 366–410 https://doi.org/10.5465/annals.2018.0174.

Khan, A., F. T. N. Gabralla, and A. Musa, 'Online Food Delivery Platforms and the Emerging Entrepreneurial Culture in Africa', *African Journal of Business Management*, 14.8 (2020): 213–22 https://doi.org/10.5897/AJBM2020.9053.

Langlois, G., and G. Elmer. "The Research Politics of Social Media Platforms." *Culture Machine*, 14 (2013): http://culturemachine.net/index.php/cm/article/view/505/531.

Le Ludec, C., E. Wahal, A. Casilli, and P. Tubaro. "Quel statut pour les petits doigts de l'intelligence artificielle ? Présent et perspectives du micro-travail en France." *Les Mondes Du Travail*, 24–25 (2020): 99–112.

Leighton, T. G., 'The Challenges of Algorithms in the Gig Economy', *Communications of the ACM*, 62.11 (2019): 56–65 https://doi.org/10.1145/3357239.

Levi, M. L., 'Los Modelos de Negocios y la Economía Colaborativa', *Revista Innovación y Ciencia*, 21.3 (2018): 38–44.

Liem, M. C., J. Zhang, and Y. Tao, 'Fairness and Transparency in Algorithmic Work Distribution: A Review', *AI and Ethics*, 2.3 (2022): 155–72 https://doi. org/10.1007/s43681-022-00124-6.

Lobato, R. *Netflix Nations: The Geography of Digital Distribution* (NYU Press, 2019).

Lopez Galvez, D., 'Digital Labor Platforms in Latin America: A Case Study of Delivery Workers in Quito', *Global Labor Journal*, 11.1 (2020): 58–73.

Lord, C., O. Bates, A. Friday, F. McLeod, T. Cherrett, A. Martinez-Sykora, and A. Oakey. "The Sustainability of the Gig Economy Food Delivery System (Deliveroo, UberEATS and Just-Eat): Histories and Futures of Rebound, Lock-in and Path Dependency." *International Journal of Sustainable Transportation*, 17.5 (2023): 490–502. https://doi.org/10.1080/15568318.2022.2066583.

Markus, M. L. "Datification, Organizational Strategy, and IS Research: What's the Score?" *The Journal of Strategic Information Systems*, 26.3 (2017): 233–241. https://doi.org/10.1016/j.jsis.2017.08.003.

Martínez Fernández, G., *Trabajo En Plataformas Digitales: Transformación, Regulación y Alternativas* (Editorial Trotta, 2020).

Maya, N., T. Quevedo, D. Carrión, and P. Sánchez. *Hacia una caracterización de las y los repartidores y de la economía de plataformas en Ecuador* (Rosa Luxemburg Stiftung, 2022).

McClain, D. C., and K. A. Posner, 'Worker Experiences on Food Delivery Platforms: Organizational Effects of Digital Intermediation', *Journal of Management Studies*, 57.3 (2020): 125–38 https://doi.org/10.1111/joms.12542.

McKinsey, D., *El Auge de Los Negocios de Plataforma En América Latina* (McKinsey Global Institute, 2022).

Meijerink, J., and T. Bondarouk. "The Duality of Algorithmic Management: Toward a Research Agenda on HRM Algorithms, Autonomy and Value Creation." *Human Resource Management Review*, 33.1 (2023): 100876. https://doi.org/10.1016/j. hrmr.2021.100876.

Mohlmann, M., and L. Zalmanson. "Hands on the Wheel: Navigating Algorithmic Management and Uber Drivers' Autonomy." *ICIS 2017 Proceedings*, December 10, 2017. https://aisel.aisnet.org/icis2017/DigitalPlatforms/Presentations/3.

Montfort, N., and I. Bogost. *Racing the Beam: The Atari Video Computer System* (MIT Press, 2009).

OECD, 'The Future of Work: How Digital Platforms Are Shaping Labour Markets', *OECD Employment Outlook 2019*, 2019 https://doi.org/10.1787/9ee00155-en.

Oestreicher-Singer, G., and A. Sundararajan. "Recommendation Networks and the Long Tail of Electronic Commerce." *MIS Quarterly*, 36.1 (2012): 65–83. https://doi.org/10.2307/41410406.

OIT. *El trabajo en las plataformas digitales de reparto y transporte en Ecuador* (International Labour Organization, 2022).

Pang, H. S., 'The Ethical Implications of Algorithmic Management on Gig Workers', *AI & Society*, 36.1 (2021): 45–56 https://doi.org/10.1007/s00146-020-01035-y.

Parent-Rocheleau, X., and S. K. Parker. "Algorithms as Work Designers: How Algorithmic Management Influences the Design of Jobs." *Human Resource Management Review*, 32 (2022): 1–17. https://doi.org/10.1016/j.hrmr.2021.100838.

Plantin, J.-C., C. Lagoze, P. N. Edwards, and C. Sandvig. "Infrastructure Studies Meet Platform Studies in the Age of Google and Facebook." *New Media & Society*, 20.1 (2018): 293–310. https://doi.org/10.1177/1461444816661553.

Posenato, A., and D. Parker, 'Digital Platforms as Drivers of Economic Innovation', *Journal of Economic Perspectives*, 34.2 (2020): 28–45.

Rosenblat, L., *Uberland: How Algorithms Are Rewriting the Rules of Work* (University of California Press, 2018).

Rosenblat, A., and L. Stark. "Algorithmic Labor and Information Asymmetries: A Case Study of Uber's Drivers." *International Journal of Communication*, 10 (2016): 27. https://doi.org/10.2139/ssrn.2686227.

Schildt, H. "Big Data and Organizational Design – The Brave New World of Algorithmic Management and Computer Augmented Transparency." *Innovation*, 19.1 (2017): 23–30. https://doi.org/10.1080/14479338.2016.1252043.

Scholz, R., and S. Schneider, 'Workers' Voice in the Digital Economy: Co-Determination and Representation', *Industrial Relations Journal*, 49.4 (2018): 312–28 https://doi.org/10.1111/irj.12224.

Srnicek, A., *Platform Capitalism* (Polity Press, 2017).

Stein, C. 'Algorithmic Cultures and the Futures of Work', *Global Media Journal*, 19.36 (2019): 1–12.

Stiegler, B. "Questions de pharmacologie générale. Il n'y a pas de simple pharmakon." *Psychotropes*, 13.3–4 (2007): 27–54. https://doi.org/10.3917/psyt.133.0027.

Sundararajan, J., *The Sharing Economy: The End of Employment and the Rise of Crowd-Based Capitalism* (MIT Press, 2016).

Timko, P., and R. van Melik. "Being a Deliveroo Rider: Practices of Platform Labor in Nijmegen and Berlin." *Journal of Contemporary Ethnography*, 50.4 (2021): 497–523. https://doi.org/10.1177/0891241621994670.

Tubaro, P., A. Casilli, and M. Coville. "The Trainer, the Verifier, the Imitator: Three Ways in Which Human Platform Workers Support Artificial Intelligence." *Big Data & Society*, 7.1 (2020): 2053951720919776. https://doi.org/10.1177/2053951720919776.

Tufekci, Z. "Algorithmic Harms Beyond Facebook and Google: Emergent Challenges of Computational Agency." *Colorado Technology Law Journal*, 13.2 (2015): 203–218.

Vallejos Rivero, O. "Control Mechanisms of Uber Platform over Its Associate Drivers in the Metropolitan Region of Chile." *Cuhso-Cultura-Hombre-Sociedad*, 31.1 (2021): 391–416. https://doi.org/10.7770/CUHSO-V31N1.ART2412.

van Dijck, J. *The Culture of Connectivity: A Critical History of Social Media* (Oxford University Press, 2013).

van Doorn, T. J., 'Stealing Time: Labor, Capital, and the Temporalities of the Gig Economy', *New Media & Society*, 19.2 (2017): 175–94 https://doi.org/10.1177/1461444816654346.

van Doorn, N., and D. Vijay. "Gig Work as Migrant Work: The Platformization of Migration Infrastructure." *Environment and Planning A: Economy and Space*, 0308518X211065049 (2021): https://doi.org/10.1177/0308518X211065049.

Vásconez, L. "Juez Confirmó Inocencia de Ciudadanos Venezolanos Detenidos en el Paro." *El Comercio*, January 21, 2020. https://www.elcomercio.com/actualidad/seguridad/juez-inocencia-venezolanos-detenidos-paro.html.

Webster, S., and L. Williams, 'Social Protection in the Digital Age', *ILO Working Paper Series*, 2020.

Woodcock, J., and C. Cant. "Platform Worker Organising at Deliveroo in the UK: From Wildcat Strikes to Building Power." *Journal of Labor and Society*, 25.2 (2022): 220–236. https://doi.org/10.1163/24714607-BjA10050.

Woodcock, J., and M. Graham. *The Gig Economy: A Critical Introduction* (Polity Press, 2019).

Zarsky, T. "The Trouble with Algorithmic Decisions: An Analytic Road Map to Examine Efficiency and Fairness in Automated and Opaque Decision Making." *Science, Technology, & Human Values*, 41.1 (2016): 118–132. https://doi.org/10.1177/0162243915605575.

Zysman, Y., and A. Kenney, 'The Next Phase in the Digital Economy: Platforms, Apps, and the Growth of Contingent Work', *Journal of Industrial Relations*, 58.1 (2016): 5–25 https://doi.org/10.1177/0022185615624011.

3

DIGITAL KNOWLEDGE FROM BELOW: LOW-SKILLED LABOUR MIGRATION TO THE GULF COUNTRIES AND TECHNOLOGY ADOPTION IN INDIA

Javed Mohammad Alam

AFTER TWO YEARS OF HARD WORK IN THE GULF, 51-YEAR-OLD SHAFIQ WAS finally returning home. His luggage included dates and perfumes, but the most precious items he carried were a mobile phone, a TV, and a laptop. In his view, these devices were not merely material rewards and symbols of his labour but tools that would help his family climb the social ladder and enter a world they had been excluded from so far. However, merely owning these devices was just the first step. The mobile phone, an iPhone of a previous generation, promised social prestige but also live communication with relatives and friends scattered in multiple Indian states and abroad, providing a vital link to the world beyond the village. The TV, a refurbished model from a reputable brand, had been said by the seller to offer reliable performance. It would become a source of entertainment and information, providing knowledge and the opportunity to make Shafiq's home a sought-after place by the neighbours and close family. For many in Shafiq's village, a TV of this quality was a luxury, and its presence

FIG. 3.1 Indian labourers at Shamkha 19, Abu Dhabi, 2024 (Parvez Alam)

in his home would mark a significant step towards improved access to media and news. The laptop held opportunities for learning and accessing information online – even though it was not yet fully clear what its use would be. Intermittent power supply and internet connectivity issues in their rural setting were still presenting significant challenges, but there would be ways to overcome these problems. The laptop was to become particularly important for Shafiq's eldest son, who needed it for learning new skills, as repeated on cardboard signs planted along the road going from the village to the closest city. With access to online courses and tutorials, he would turn the laptop into a tool for self-education and overcome the limitations of the public school he was attending. This would then bridge the gap between Shafiq's rural upbringing and the demands of a technology-driven economy; it would also fulfil Shafiq's wife's claims – although herself uneducated, she was prioritising the children's education, understanding its importance for their future. While migrant families like Shafiq's were starting to integrate these technologies, many non-migrant families in his village still lacked such devices, highlighting a significant and persistent divide in access to technology. This divide was not just about access but also about the ability to use technology effectively. Shafiq's return would initiate a shift for the family to engage with ongoing digital transformations.

This story, based on discussions with one migrant labourer and his family in northern India, is not unique. Across India, particularly among marginalised communities, returning migrants are bringing back more than material wealth – they are carrying with them tools and knowledge that facilitate economic integration, enhance educational opportunities, and strengthen community adaptability. Scheduled Castes (SC), Scheduled Tribes (ST), and Other Backward Classes (OBC) in India represent historically marginalised communities facing systemic discrimination on various fronts, including education, employment, and social status. Despite constitutional safeguards, these groups continue to grapple with socioeconomic challenges. According to National Sample Survey Office (NSSO) data, the literacy rate among SCs was 66.1% compared to the national average of 73% in 2017–18, highlighting educational disparities. STs experience even greater educational marginalisation, with a literacy rate of 58.9%. Muslims, especially the Pasmanda[1] community – a term encompassing Dalit and economically disadvantaged Muslims and designating a marginalised subset of the Indian Muslim community – face similar, if not heightened, levels of discrimination. The Sachar Committee Report (2006) found that the socioeconomic status of Muslims in India is comparable to, or worse than, SCs and STs. Pasmanda Muslims often fall into the lowest brackets of the social hierarchy, enduring economic deprivation and a lack of educational opportunities. This marginalisation is compounded by the pervasive burden of communalism. The rise of majoritarian political ideologies, particularly under the current regime, has exacerbated the discrimination against Muslims. The 2011 Census of India already highlighted that unemployment rates among Muslims in Uttar Pradesh were higher than the national average, underscoring the paucity of opportunities at home. Pasmanda Muslims in Uttar Pradesh increasingly migrate to Gulf countries in pursuit of work opportunities and to overcome socioeconomic deprivation and educational disadvantages (Bhatty 1996). This migration trend mirrors the earlier patterns observed in Kerala, where economic necessity, combined with limited local opportunities, prompted large-scale emigration (Chandramalla 2022).

Out of the approximately 9 million Indian workers in the Gulf region, a staggering 90% are employed in low- and semi-skilled positions (ILO 2018). Despite the importance and complexity of this phenomenon, it has often been analysed primarily through the lens of migrants' working conditions. Their significant contributions to knowledge exchange and technological globalisation have not received adequate attention. Studies of low-skilled migration to Gulf countries have mainly focused on labour-market segmentation, exploitation and the challenges faced by migrants in terms of working conditions, wages and social integration (Breman 2019; Chandramalla 2022; Khan and Harroff-Tavel 2011; Srivastava 2021). One explanation for this situation is the frequent association of 'knowledge' and 'technology' with high-skilled work. Low-skilled labour migrants, concentrated in labour-intensive industries, face structural barriers to participation in technological innovation; their engagement with technology primarily revolves around the use of digital platforms for communication and remittance transfers. In contrast, skilled migration has long been studied through the 'brain drain' lens, which emphasises the loss of skilled human capital in developing countries due to emigration (Bhandari 2019; Dhar and Bhagat 2021). Following research on brain drain, attention shifted to the concept of 'brain circulation', recognising the potential benefits of high-skilled migration through the transfer of knowledge and technology back to migrants' home countries (Tsai 2020). This new perspective highlights the role of highly skilled and globally mobile individuals in fostering the growth of digital industries in the Global South (Saxenian 2005; Saxenian and Hsu 2001). These individuals often have strong connections with professionals and institutions in their destination countries, allowing them to access valuable resources and information (Lo et al. 2019; Pollozek et al. 2021; Rajan and Kumar 2020). Their transnational networks become conduits for the dissemination of technological knowledge as high-skilled migrants bring back new ideas, practices, and contacts to their home countries (Chorev and Ball 2022; Khadria 2006; Lo et al. 2019; Williams 2007). For example, IT specialists and software engineers of Indian origin have played a crucial role in India's technological development, exemplified by the rise of Bangalore as the Silicon Valley of the Global South (Upadhya 2016; Wise 2022; Zweig et al. 2021). The focus on high-skilled migrations has, therefore, tended to overshadow the knowledge and technology flows happening through low-skilled migrations.

In order to overcome the lack in the literature and understand what kind of knowledge and technological know-how circulate through low-skilled migration, one has to consider that low-skilled migration to the Gulf countries involves different technology-dissemination and knowledge-transfer dynamics. This study is set against the backdrop of rapid technological globalisation and migration trends between India and the Gulf countries. It investigates how low-skilled labour migrants contribute to shaping the digital and technological landscapes of their home communities. The primary research questions include the following: What are the conditions and contexts in which migrants to Gulf countries acquire digital devices? How do these migrants adopt and use technology? What impact does their technological engagement have on their families and communities in India? Low-skilled migrants actively participate in the globalisation of digital technologies, influencing technological practices and expectations in their home country, India. The chapter explores how migrants, as social actors, redefine their relationships through technology, influencing both their immediate and extended social environments.

From October 2022 to February 2023, I conducted semi-structured interviews to delve into the experiences of Pasmanda Muslim migrant labourers in Saudi Arabia, the United Arab Emirates (UAE), Kuwait, and Qatar. The sample consisted of 31 participants, 21 of whom were interviewed online due to their current residence in these Gulf countries. The virtual format allowed for real-time conversations despite geographical distance, capturing the immediacy of their experiences. Additionally, I carried out in-person interviews with the family members of 10 migrants in a village in the Deoria district of Uttar Pradesh, India. This village, predominantly Bhojpuri-speaking, has a mixed population of Muslim and Hindu families with low educational levels and few white-collar workers. About 60 Pasmanda Muslim families live here, most of them with few or no land holdings. Conducting interviews in the village provided rich, contextual insights into the local socioeconomic impacts of migration. The interviews were conducted in a mix of Bhojpuri and Hindi. These face-to-face interactions enabled a deeper connection and understanding of how migration shapes the daily lives and future prospects of these families.

To understand the intricate dynamics of digital technology globalisation through labour migration, this chapter adopts an interdisciplinary approach rooted in science and technology studies (STS).

The theoretical framework provided by Winner (1980) and Bowker and Star (2000) sets the foundation for analysing how digital technologies are integrated into social contexts, which is crucial for understanding the experiences of low-skilled labour migrants. Decolonising methodologies (Smith 2021) ensure that the research accounts for the power dynamics and colonial histories that shape technology adoption among migrants. Anderson and Adams (2008) highlight the integration of local and global knowledge systems, which is essential for understanding how Indian migrant workers adopt technologies in the Gulf, emphasising the interplay between global flows and local adaptations. Similarly, Beaudevin and Pordié (2016) illustrate the adaptation of biomedical technologies in diverse cultural contexts, offering parallels to digital technology use among migrants. Furthermore, Behrends, Park, and Rottenburg (2014) explore the translation of technologies across contexts, which helps us understanding the interaction between digital tools and local practices among migrant workers. Adding to this, Dumoulin Kervran, Kleiche-Dray, and Quet (2018) stress the importance of engaging Global South perspectives by showcasing unique interactions with digital technologies. In line with these views, Mavhunga (2017, 2014) discusses everyday innovation and creative adaptation, which is relevant to the experiences of Indian migrants. Rottenburg, Schräpel, and Duclos (2012) reflect on the relocation of science and technology within the Global South, involving complex sociocultural adaptations. Twagira (2020) and Von Schnitzler (2013) emphasise the materiality of politics and infrastructure, demonstrating how technologies shape and reflect the social realities of migrant labourers. By drawing on these interconnected STS perspectives, this chapter provides a comprehensive understanding of how digital technologies are adopted and adapted by low-skilled labour migrants. The integration of these insights highlights the critical role of local contexts, cultural practices, and power dynamics in shaping technology use and adaptation, ensuring a grounded analysis in both theoretical and empirical considerations.

As an STS researcher and a member of the Pasmanda Muslim community, my positionality significantly shaped this research. Sharing a common background with the participants facilitated trust and openness during the interviews. Informed consent was obtained from all participants, ensuring they understood the purpose of the research and their rights. Confidentiality and anonymity were strictly maintained to protect the participants' identities. This methodological approach, anchored in STS and enriched by my positionality as a Pasmanda Muslim researcher, aims at providing a nuanced understanding of the interplay between digital technology and labour migration. It highlights the social structures, practices, and outcomes experienced by migrant workers and their families, offering valuable insights into the broader implications of globalisation and technological change for marginalised communities in Global South countries.

The chapter is structured to provide a comprehensive analysis of low-skilled labour migration from India to the Gulf countries and its impact on knowledge circulation and technological adoption. Following the introduction, the chapter is divided into four sections and a conclusion. The first section explores the economic drivers of migration and the role of social networks in facilitating this movement. The second section delves into how migrants use digital technology to maintain connections with their families and foster economic interactions. The third section examines the adoption and use of technology by migrants and their families in a broader perspective. The fourth section focuses on the grassroots knowledge circulation facilitated by migrants and their contributions to local innovation and entrepreneurship. The chapter concludes by underscoring the significant contributions of low-skilled migrants to knowledge exchange and technological globalisation, advocating for a more nuanced understanding of their impact on socioeconomic landscapes in their home countries.

LOW-SKILLED MIGRANTS' PATHWAYS TO THE GULF COUNTRIES

Low-skilled migrants are often driven by economic factors to seek employment opportunities in the Gulf countries (Azhar 2016; Bélanger and Rahman 2013; Shah and Menon 1999; Sultana and Fatima 2017). These individuals come from marginalised backgrounds and face limited prospects in their home countries due to low wages, a lack of job security, and inadequate living conditions. The promise of higher wages and better economic prospects in the Gulf countries serves as a strong incentive for them to migrate. Economic drivers play a crucial role in informing the decision-making of low-skilled migrants, who aspire to improve their standard of living and provide financial support to their families (De Haas 2011). However, the journey itself is fraught with financial difficulties. To fund their migration, migrants resort to desperate measures, such as borrowing money from relatives, selling their land, or falling prey to loan sharks, who exploit their vulnerable situation (Kamat and Crabapple 2015; Rosenberg et al. 2017). Many of the migrants come from socially disadvantaged communities among the lower rungs of the caste hierarchy. These communities have historically faced discrimination and marginalisation, leading to limited access to resources, opportunities, and social mobility (Jodhka 2001; Srinivas 2003). Thus, labour migration provides a strategy for these individuals to escape the structural constraints and systemic inequalities that persist within their home communities.

Over the years, India's low-skilled migrants have formed social networks that extend from their home villages to the cities of Gulf countries (Kathiravelu 2012; Khadria 2006). These networks are made up of people who share similar geographical roots and who have either already migrated or possess knowledge about employment opportunities and visa procedures. The tight-knit nature of these networks creates a strong sense of community and support as migrants exchange information, advice, and assistance with one another. By utilising these networks, migrants obtain vital information about potential job openings, navigate bureaucratic procedures, and form connections in foreign countries. This can include information about job opportunities, visa requirements, and cultural norms in the host country. A young man waiting for a visa, for instance, obtained the necessary information from elders: 'In my village, everyone wants to go out. The first thing we do is to get a passport as soon as possible. We all know how we can go [outside India]'. The young boys in the villages can see the change afforded by remittances, citing the examples of people who have worked in the Gulf for years. 'What is possible from Saudi money is not possible in India. People [migrants] have built houses, bought cars and bikes, sending kids to good schools. Can it be done working here [in India]?'

As regards the Gulf countries, low-skilled and semi-skilled workers typically have two avenues for entering the labour market: either through a Kafeel[2] or through recruitment agencies. Kafeel, meaning 'sponsors' or 'employers,' plays a crucial role in the lives of migrant workers in Saudi Arabia (Garratt 2021; Parreñas and Silvey 2021). The Kafala system is a sponsorship programme that regulates the entry, residency, and exit of foreign workers in the Kingdom of Saudi Arabia (Parreñas and Silvey 2021). The system has been in place for several decades, governing the legal and administrative formalities of low-skilled and semi-skilled workers, including visas, work permits, and residency permits. It imposes certain restrictions on their movements, employment opportunities, and social mobility. A migrant who had been running an independent electronic repair business for a decade explained:

> I came with the help of friends already here [in Saudi Arabia]. I worked with my friend in the beginning and later started my own repair shop. I have a good relationship with my Kafeel, and he allows me to work independently, but I have to pay 1,000 Saudi riyals (approximately 267 USD) every month. He [the Kafeel] doesn't care how much I make from the business [a repair shop].

The migrants' experience with their Kafeel is subjective, and their arrangements depend on their personal relationships and mutual benefits. The interviewed shopkeeper provided further insight into the role of the Kafeel in the setup of his business:

> My Kafeel is a government employee, and he can't run the business under his name, so the shop is under his wife's name. In Saudi Arabia, the [non-working]

housewife gets the unemployment benefits from the government, which is very much what I pay, but since she has a business in her name, she doesn't get the benefits from the government.

When prompted, he also explained why his Kafeel kept a business that is not profitable: 'In the end, it is an asset. He [the Kafeel] can take charge of it, expand, and run himself after his retirement'. Migrants from a particular village or locality in India usually go to the same city in Saudi Arabia, often under one Kafeel. The reason for this is the presence of a tight-knit network, usually composed of relatives or fellow villagers who have migrated and can secure sponsorship from a Kafeel under their own name. One worker I interviewed had three generations and seven members of his extended family who worked in the same Kafeel system in Tabuk, Saudi Arabia:

My father came first, and then I came. Now, a lot of my relatives and cousins are here. It's much easier for those who already have relations with Kafeels. We just pay them [the Kafeels] all expenses for each worker. Here, a lot of people from my district [in India] and nearby places.

Not all migrants under the Kafeel system, however, are free to work independently. The visa issued under the Kafeel system is to fulfil the particular needs of the Kafeel. Hence, migrants cannot change jobs or leave the country without their sponsor's permission. A driver went through a difficult experience with his Kafeel: 'I thought that it would be just driving, but I had to do a lot of other household work besides driving. I came back [to India] and had to arrange another visa'. The Kafeel has a significant amount of power and control over the workers, with the ability to withhold wages, confiscate passports, and restrict workers' movements, making it difficult for them to leave the country or find alternative employment. One migrant working as a construction labourer in the UAE never returned to Saudi Arabia after his first trip: 'When I first arrived, I wasn't aware of the purpose of my visa. They sent me to a goat farm. I stayed there for two years'. The Kafala system has been criticised by human rights organisations and labour unions for its potential to abuse and exploit foreign workers (Malaeb 2015). Nevertheless, it has provided a pathway for migrants to enter and find employment through established networks of fellow migrants.

Recruitment agencies, on the other hand, are private companies that facilitate the recruitment and placement of workers in Gulf countries. They often have agreements with employers to provide workers for specific industries, such as construction, hospitality, and domestic work (Fernandez 2014). They are responsible for screening and selecting workers, arranging visas and travel, and providing support to workers upon arrival in the host country. Recruitment agencies have offices in both India and the Gulf countries. They train the workers going to the Gulf countries for the first time, as described by a worker in Abu Dhabi: 'They kept us at the office [in the nearest district] to train for the work and how to board a plane and navigate at the airport on arrival'. One of the primary issues associated with recruitment agencies is the high recruitment fees they charge: 'It's expensive, but there is no other way. If you want to go out [to Gulf countries], you have to pay'. Migrant workers continue to face significant barriers at both ends of their transnational migration. A worker in Oman told me about the Indian government's ECR (Emigration Check Required):[3] 'I don't see the point, how it helps us. We have to go for a stamp before we come here [to Gulf countries] for no reason'. ECR passport holders are workers with a lower level of education, such as those who have not earned the Secondary School Certificate (SSC) or passed the Secondary School Examination (SSE), typically taken around the age of 15–16, and who are seeking low-skilled or semi-skilled employment abroad (Zachariah and Rajan 2016). In 2015, the Indian government introduced the eMigrate[4] system to facilitate the online registration of workers seeking employment abroad and to provide greater transparency and accountability in the recruitment process.

The migration process is a challenging endeavour for individuals from marginalised backgrounds, who often lack the financial means to fund their journey and initially do not know how to use the technological devices that can facilitate any migration experience. These migrants rely on their social networks to gather information and make informed decisions about their destination, job prospects, and living conditions. While skilled migrants can navigate the digitised procedures of high-income countries, labour migrants face additional barriers in accessing migration pathways and often resort to 'illicit' routes. These diverging experiences highlight the complex interplay of socioeconomic factors in shaping migration pathways and suggest equally different experiences with the adoption of technology.

MIGRANTS' SOCIAL NETWORKS AND TECHNOLOGICAL USES IN THE GULF COUNTRIES

The migrant workers' social lives in the Gulf first rely on the connections made through their worksite and on their housing conditions. In the case of employment within a company, the organisation is typically responsible for arranging accommodation for the migrants, leaving them with little to no say in the matter. A construction worker in the UAE described the living arrangements organised by the company: 'This is like a hostel, and bunk beds are all across the room. One bed is allotted and nothing else, but it is just to sleep; it's not like home'. The experience of sharing accommodation among migrants, combined with the pervasive feeling of uncertainty in their lives, cultivates a deep emotional connection to their home and family. The shared accommodation in which migrant workers stay fosters a sense of camaraderie among workers, who all have a temporary status in a foreign country. A migrant in Saudi Arabia shared his experience:

> People live here [in the Gulf] in shared accommodation to save money. Eight to ten people sleep in this room, but I haven't met a few of them. Everyone has their timings [for work] [...] Every now and then, new people come and go. It is like this here.

Migrants under the Kafala system have developed an understanding of how to negotiate rules, which allows them to share private accommodation based on what is more convenient for them and on their social connections within the network of fellow migrants, often comprising people from the same geographical location in India. A plumber working in Jeddah, Saudi Arabia, told me about his arrangement with the Kafeel: 'We are supposed to live with Kafeel or wherever the work is, but people doing business live independently. But those who work in farms get a place to live'.

In challenging situations, migrants rely on a broader social network at home for support and assistance. During the COVID-19 lockdown in India, a significant number of migrants were compelled to return home and encountered various challenges. However, their support network played a crucial role in helping them overcome these difficulties and maintain their stability. As a young migrant in Saudi Arabia explained: 'I came back suddenly and didn't get my pay, but we keep in touch with [each] other, borrowed from friends and relatives, we managed'.

In that regard, migrants in Gulf countries are led to adopt and leverage digital technology to maintain connections with their families and communities back home. Technological advances, including the affordability of smartphones and improved internet connectivity, have deeply transformed communication for migrants, enabling them to stay connected with their families and to support them by sending remittances, which are often these families' primary sources of income. The rise of digital platforms, such as social media, messaging apps, and online marketplaces, has further facilitated the creation of local networks that connect transnational migrants, fostering social and economic interactions within these communities. These digital tools have become integral to shaping the transnational network of migrants, enabling them to maintain relationships, share information, and engage in economic activities across borders. Communication technology has helped them to maintain a constant communication channel with their relatives and social circle back home, influencing their consumption choices in the Gulf region. A shop owner living in Saudi Arabia since 2000 has seen a drastic change:

> When I arrived, there was no WhatsApp like today. People used to send things and exchange letters with those who went on vacation until they went home. [….] Now we can buy online from here. We buy things as demanded by kids. [….] I took a flat TV home when I first went back [to India]. No one in my village has a flat TV. Now everything is available in India, but things here are of better quality and even a second-hand look like new.

The selection of items migrants purchase is influenced by both trends in the Gulf countries and what they desire in their home country. While interviewing this migrant's family in India, I noticed older versions of iPhones that remain out of reach for most Indians. In another instance, an electronic repair shop owner had been sending home useful items via courier services: 'It used to cost about 200 Riyals [approximately 53 USD] for about 50 kg. I have sent a TV, printers, and other items. Now most people buy phones and laptops'. Migrants working with a company may not have the freedom to buy things, as explained by a migrant in Dubai: 'We buy things which are necessary or when we go home. Before the flight, we go to the market and buy things for our home and kids'. Migrants and their families back home are thus exposed to the latest technology arriving from the Global North in Gulf countries. The longstanding perception of the superior quality of goods arriving with migrants still resonates with this migrant's family: 'I have a ten-year-old TV, and it still works better than new [Indian] ones'.

During my interview with the migrant families, I observed a notable trend in which they possessed multiple smartphones, often from expensive brands. These smartphones were brought home by the migrants, who bought them as used devices. This makes the migrants' families stand out, considering the higher prices of smartphones in India, which makes them unaffordable for labourers. This is also emblematic of the second generation's growing level of education and aspiration to keep up with technological trends. Migrants in Gulf countries rely on financial and technological resources to build a sustainable life for their families. A driver in Abu Dhabi sent home two laptops and one smartphone the previous year: 'I have sent two laptops and one Samsung [smartphone] for online education. I don't want them to say that they [the kids] can't learn because they don't have a laptop or mobile'. Low-skilled migrants rely on their family members to keep up with policy changes by the Indian government so as to maintain their democratic rights in their home country. The changing landscape of digitisation in India has created a two-way stream of information. A migrant in Dubai who goes home every two years told me the following:

> There is so much work for us to do at home. The government is always changing rules [....] so, to avoid any glitches, I take extra care. I have an

Aadhaar[5], PAN card, voter ID, and ration card in my name. I ask my son to update and make changes.

The technology brought back from the Gulf region by migrants for their family members often includes used or previously owned devices, which migrants refer to as 'second-hand devices'. Although they have already been used in rural villages, these devices offer the family members a novel experience that would otherwise be inaccessible to them. Within their social and financial group, they are the first to use these preowned devices – fully functional, previously owned devices purchased by a second user at a significantly lower price than their original cost. This enables the migrants and their families to become comfortable with multiple devices and mobile applications, be it for online banking, entertainment, learning, or communication.

More generally, the inflow of remittances has a significant impact on the economic wellbeing of families who have been relying on daily wages or marginal agriculture. The added household income allows for an increase in daily expenses (food, clothing, and cultural activities) and investment in education, healthcare, private transportation, and digital communication devices. Once the workers are in a stable enough position and have learned to navigate between jobs, their focus shifts to building concrete houses, paying for private education, or buying a refrigerator, a TV, and a motorbike. Workers in Gulf countries share similar goals, as a migrant who had been working with a construction company in the UAE for the last five years explained:

> If I was working at home [India], I would not have built the concrete house. All of us are sending money to take care of the family. What else are we here for? [...] Children are getting the education; they will find a good job [in India].

This situation also changes educational aspirations. Over the previous five years, 14 sons of migrants from the village where I carried out interviews had completed their engineering degrees. One migrant from the village told me what made this possible:

> We came here [in Saudi Arabia], which made it possible for us to send them to school. No one could think of it ten years back. My generation only wanted to find work. Some of us haven't gone to school.

The fact that all 14 engineering graduates from the migrant village were boys, however, shows the inherent issues surrounding the gendered allocation of resources in the community (Amirtham and Kumar 2023; Antonio and Tuffley 2014; Gupta and Sharma 2003). It is not a lack of access to education but cultural norms that prevent the daughters of migrants from opting for engineering degrees. Many Indian families prioritise investing in the education of sons over that of daughters. After being unable to find a job for three years after graduation, an engineer from the village chose to join his father in Saudi Arabia. The father and son are now both under the same Kafeel, with the father having a school certificate and the son holding an engineering degree: 'There are no jobs. [...] He [the son] was just waiting for an opportunity, so I asked my Kafeel for one more visa. He will help me with the shop'.

Living and working in the Gulf countries is a transformative experience for the migrants. It changes their use of digital technology, which suddenly becomes an essential tool to remain connected to those back home. The latter – the children, wives, and parents of the migrant workers – are also affected by this experience, as the following section demonstrates by highlighting the adoption and use of technology among migrant families in Indian villages.

THE ADOPTION AND USE OF TECHNOLOGY BY MIGRANTS AND THEIR FAMILIES

In India, the ability to manage uncertainty and navigate challenging circumstances is influenced by a wide range of overlapping factors based on income group, social class, gender, and caste (Guérin 2014). The wellbeing and livelihood of poor families are intricately tied to the support they receive from government inclusion policies, assistance programs, and remittances. These families heavily depend on these forms of assistance to meet their basic needs and sustain their daily lives. These policies aim to address the financial and social challenges faced by groups at the margin, ensuring they have a safety net and opportunities for upward mobility. In that sense, they complement the remittances sent by migrant workers back to their families. Remittances often form the backbone of the family's financial stability and play a significant role in improving their quality of life. However, migrants' take-up of government aid programs is determined by everyday practices, the attitudes of local government bodies, and social and economic factors. They do not perceive these policies as a one-stop solution; instead, they interpret them in their unique ways to optimise their options. A migrant worker living in Saudi Arabia for the last eight years, who was a beneficiary of government schemes, discussed the use of subsidies:

> Whatever government gives is not enough, but it does help. [...] I sent my son to a private school because I earned more than what I used to in India; otherwise, it was not possible only with food subsidies.

Against the backdrop of rapid digitisation, all of the Indian government's inclusion schemes (social, financial, health, and education) have moved online and work with citizens' virtual IDs (Martin 2021; Singh 2019). This was made possible by integrating three technologies, namely bank accounts, mobile numbers, and biometric identification (Aadhaar), to create a verified virtual presence for each eligible citizen (Rao and Nair 2019). The move towards the 'platformisation'[6] of governance, away from the earlier mix of physical and virtual presence, has received considerable criticism from scholars for exacerbating the inadequacy resulting from a system that privileges individuals with specific skills, resources, and educational backgrounds (Bowker and Star 2000; Masiero 2017, 2018). However, migrants are among those who have already embraced digitisation and own smartphones, allowing them to take advantage of the digital transformation. A migrant in Tabuk, Saudi Arabia, saw the transformation as 'necessary' for him:

> Now, I can monitor and check expiries and dates of benefits. I talk to the village head regarding certificates, and I apply online. [...] I go to my friend here [in Tabuk] if I need anything for my family. It is necessary for me.

Remittances help with the acquisition of resources, but migrants and their families decide on the prioritisation of goals. The women of the family have gained some autonomy as they manage the house in the absence of the men. They decide on which hospital to go to, where to send the children for a good education, and whether they really need transportation or a new mobile phone. 'He comes home after two years and only for one or two months, so I have to go out and do things', said the wife of a migrant in Kuwait. However, this autonomy is not absolute: face-to-face conversations on video calls have enabled men to take much interest in family matters in a way that was not possible earlier. This example is important for recognising that digital devices can also perpetuate existing social and cultural norms, particularly concerning gender dynamics (Nakamura 2007; Shevinksy 2015). Gender norms are deeply entrenched in Indian societies, and digital devices such as smartphones and laptops and social media platforms have become tools through which these norms are both conveyed and reproduced. Migrants use digital spaces to assert control and power, often perpetuating patriarchal values and reinforcing traditional gender roles.

The daughters of migrants are able to access the schools and colleges within 15 kilometres of their homes by scooter. Among four girls of a migrant family who had managed to graduate, one had finished a nursing course: 'We had cycled for the school, but college is fifteen kilometres from here, so we insisted on buying a scooter'. During lockdown, schools shifted to online teaching, which affected millions of families across India. Many families chose to wait for the COVID-19 pandemic to be over, while some managed to buy mobile phones for the online classes (Jena 2020). The families of migrants I interviewed already had at least one smartphone at home, which they used for online lessons. One of them described the situation: 'We already had one smartphone. I talked to my husband [who lives in the UAE], and he brought one more phone with him so both of my children could attend the classes'. Among the vectors of financial inclusion, smartphone-enabled money transfer (or mobile money) tools have encouraged a culture of online shopping, information access, and keeping up with the trends across different mobile applications.

A migrant running an electronic repair shop in Tabuk, Saudi Arabia, told me: 'I am very busy. The working hours are different in Tabuk. By the time I close the shop, it is after midnight in India. So, we talk while I am working. I decide if something is important; otherwise, my wife and my kids will take care of the things'. The families of migrants in India are also taking part in less traditional dimensions of consumption in the villages. They opt for city-based services over local shops because they are influenced by the perception of cities as key places for finding the latest fashion, as the young daughter of a migrant explained: 'I don't shop in the town. It takes a whole day to go just for shopping, but I buy from the city, they have design and quality. […] Here only a few shops are available'. Young, educated daughters of migrants are adding gendered dimensions to the functioning of everyday life by exercising choices afforded by remittances.

For migrant workers, sending remittances has become easier with the growing use of mobile phones to transfer funds directly to family accounts in India. Whereas formerly remittances were transmitted to the family account through traditional banking platforms, payments and transfers now take place seamlessly through mobile money or on fintech platforms in India. (Bhaskar 2013). The funds are securely held by banks, and financial transactions can be carried out through either traditional banking channels or fintech platforms. Despite the uneven distribution of ATMs in rural India, cash withdrawals from ATMs are a popular channel for obtaining money. However, digital transactions have been gaining prominence in rural India due to the popular claim that they are cheap, convenient, and fast, with added multidimensional security to avoid phishing and fraud. Nevertheless, a migrant in the UAE told me that he learned about digital transactions before they became popular in his village: 'Here [in the UAE], digital systems have been available for a long time now. When I came here [in the UAE], I had to learn how to use and operate my mobile, but now I do everything on my mobile. All in it. […] Now, everything is digital at home [in India] too'. The sociotechnical relationship extends to the network of migrants, which is built on mutual interests, regional ties, and willingness to assist those in need. The network of migrant workers, early access to technology, and the inflow of remittances from the Gulf countries have enabled family members in India to become part of the formal circulation of techno-financial resources.

LOW-SKILLED LABOUR MIGRATION AND KNOWLEDGE CIRCULATION FROM BELOW

Low-skilled labour migration to Gulf countries establishes a form of digital knowledge circulation from below, wherein low-skilled migrants contribute to the transnational flow of information, skills, and resources that benefit their home communities. This grassroots-level knowledge circulation is driven by several key factors, including the migration process, the social networks formed, and the adoption of digital technology by migrants and their families.

Social networks, composed of people who share similar geographical roots, become essential sites for exchanging information, advice, and assistance. Migrants obtain vital information about job openings, navigate bureaucratic procedures, and form connections in foreign countries through these networks. Often, migrants follow a chained network, working alongside someone familiar with the job and local customs. For instance, migrants working as technicians start by assisting well-trained colleagues and learning on the job, acquiring tacit digital knowledge that is not formally taught but learned through observation and practice. This includes understanding the use of specific tools; mastering repair techniques for mobile phones, air conditioners, and washing machines; and adapting to the local work culture and language. This is particularly true for the older generation of migrants who arrived with little or no knowledge of a particular skill and had to learn whatever paid best. A 50-year-old migrant, completing his 20th year in the UAE, working at a repair shop in Bur-Dubai and additionally handling any domestic repair work, explains: 'In the beginning, it was just mechanical parts to change, and now everything that has a motherboard and chip'. He further adds how he learned advanced digital repair techniques in Dubai and trained his son to start a small electronic repair business in his town in India: 'In Dubai, I learned how to repair complex electronic devices that we never saw back home. My son also learnt it and runs a mobile repair shop at home'. A migrant in his 50s who used to work for a home appliance brand franchise for his Kafeel in Mecca, Saudi Arabia, now back in India, also explains:

I got training from the company on how to use diagnostic software for appliances, change the parts, and fix the devices. Now, I use that knowledge to fix everything from smartphones to refrigerators at my shop [in India].

Of course, knowledge circulation is never a one-way process. Some of the newer generations arriving on labour visas are educated and have diplomas in chip repair from well-known training centres in India. A 25-year-old migrant who arrived in Tabuk, Saudi Arabia, with the help of a friend, is working at a mobile repair shop owned by a Saudi citizen; he describes the differences between the devices they handle in Saudi Arabia and those used in India: 'In India, I was working as a technician where we mostly work with Chinese mobile phones, and they are cheaper and easier to repair, but here [in Tabuk, Saudi Arabia], most people have Apple and expensive Samsung, and one has to be very careful and use proper tools to open and change the chips. [...] But it is my training in India that helps me here too'. These instances underscore the role of digital knowledge circulation facilitated by social networks and informal training, contributing to both individual and community development.

In a broader perspective, the adoption of smartphones notably influences family roles and dynamics, particularly within households of migrant workers. Traditionally, major financial decisions and family resource management were the purview of the male migrant. However, their physical absence due to migration and the introduction of smartphones has led to a significant shift in decision-making power within the family. Women, often left to manage the household alone, have assumed greater responsibilities out of necessity. One woman with a school education and two young kids, eight years and four years old, described how she learnt to manage her family's finances through online banking, a skill she acquired out of necessity. She said, 'My husband used to handle all the money matters, but after he left for the Gulf, I had to learn. With a smartphone, I could manage our bank accounts, pay bills, and buy on Flipkart [an online marketplace]'. Smartphones facilitate this transition by enabling women to manage household finances, make online purchases, and access information about government schemes and policies. This transition stems from both the necessity of assuming traditional male roles and the practical utility of smartphones. At the same time, continuous communication via video calls and messaging apps allows men to remain involved in family decisions and provide remote guidance, maintaining family cohesion and support as well as reproducing certain domination mechanisms (Madianou and Miller 2013). As one migrant in Abu Dhabi explained: 'Even though I am miles away, I can still guide my wife in financial decisions through video calls'.

This digital integration creates a feedback loop in which knowledge and skills gained abroad are reintegrated into the home community, fostering local development and further enabling digital education (Vertovec 2004). For instance, a migrant bought a laptop after the pandemic in 2019 so that his children could access online educational resources for their academic performance. He describes the reasons for buying a laptop as the inability of schools to provide digital tools for his kids: 'With the laptop I sent from Dubai, my children are now able to take online classes. This would have been impossible with just the local school resources'. Similarly, the use of digital remittance platforms streamlines the process of sending money home, making it faster and more secure. A migrant in Saudi Arabia explained the changes brought by digital devices: 'Before [digital], I had to rely on money transfer agents, which took days. Now, with digital wallets, my family receives money instantly, and they can use it for urgent needs without delay'. This phenomenon underscores the role of digital tools in improving certain aspects of financial inclusion and economic participation. Migrants act as conduits of information, skills, and resources, and through their position they make digital knowledge circulate from one country to the other.

Low-skilled migrants also display significant adaptability in their use of technology, often repurposing digital tools to meet their specific needs. In some cases, migrants explained to me how they used basic technological skills learned abroad to start entrepreneurial ventures, such as mobile phone repair services or digital printing businesses in their home villages. These examples of creative adaptation illustrate how migrants actively contribute to local economies and demonstrate their agency in shaping technology to fit their circumstances. Upon returning home, they pass these skills on to family members and neighbours, often introducing technologies that were previously unfamiliar in rural areas. This – often ignored – process of knowledge transfer accelerates the adoption 'from below' of digital tools in marginalised communities, enabling greater participation in the digital economy. Through this two-way exchange, migrants contribute to digital innovation and development on both local and transnational scales.

The role of low-skilled migrants in digital knowledge circulation challenges the traditional focus on high-skilled migration (Saxenian 2005). This grassroots knowledge circulation underscores the importance of considering the lived experiences and contributions of low-skilled migrants in discussions about globalisation and development. The role of tacit knowledge, informal train-ing, and social networks highlights the complex interplay of economic, social, and technological factors in shaping migration pathways and their impacts on home communities (Di Maria and Stryszowski 2009). Migrants' exposure to new technologies and their subsequent integration into local practices foster innovation and development at the grassroots level. These migrants are not merely labourers but vital agents of socioeconomic change, facilitating a form of globalisation that is inclusive and grounded in the realities of marginalised communities. By focusing on the intersection of migration, technology, and social change, this integrated analysis advocates for a broader recognition of the contributions of low-skilled migrants. Their experiences and actions are integral to understanding the complexities of today's migration and its impact on global knowledge circulation and technological adoption.

CONCLUSION

The story of Shafiq and the broader experiences of Pasmanda Muslim migrants underscore the transformative potential of labour migration beyond mere eco-nomic gain. This migration, driven by economic necessity, not only addresses immediate financial needs but also catalyses significant sociotechnical transfor-mations in the migrants' home communities. Remittances sent back by migrants are crucial in elevating the socioeconomic status of families and enabling invest-ments in education, healthcare, and housing, thereby fostering social mobility and improving living standards (Reserve Bank of India 2018). Research by

Levitt (1998) on 'social remittances' emphasises that migrants carry more than just economic capital back to their home countries: they also transfer social norms, practices, and ideas. Digital knowledge is becoming a growing part of such phenomena. These migrants' experiences challenge the assumption that digital transformation is solely driven by top-down initiatives and highlight the importance of grassroots innovation and creative adoption in the Global South. Through the circulation of know-how gained abroad, digital inclusion takes a particular shape for marginalised communities that are often thought of as simultaneously disconnected from globalisation and technology – excluded from global technological networks. The intersection of digital transformation with other aspects of social change, such as financial inclusion and education and the expansion of social media, is evident in the lives of migrant families. The availability of digital wallets, mobile banking, and online learning platforms is expanding opportunities for economic participation and educational development in unexpected ways.

This knowledge circulation is further enhanced by the social networks formed by migrants. These networks facilitate the exchange of information, advice, and assistance, thus enabling migrants to navigate bureaucratic procedures and integrate new skills and technologies into their home communities. Migrants not only bring back devices and technical skills, but also new perspectives on their use and potential, influenced by their experiences abroad. This process underscores the contingent nature of technological adoption and the role of social groups in shaping technological outcomes (Bijker and Law 1994). This uncodified and hardly documented form of knowledge circulation is crucial for local technological adoption and adaptation. For example, migrants returning from technologically advanced regions often bring with them new skills and know-how that can spark innovation in their home communities (Vickstrom 2014). Studies such as those by Vertovec (2004) highlight that the transnational networks formed through migration serve as channels for the flow of technical knowledge and practices. These networks enable the exchange of information and skills that might not be formally recognised or codified by the state but are nonetheless vital for local development. The spread of mobile phone usage in rural Africa, facilitated by returning migrants, is a case in point (Aker and Mbiti

2010). This phenomenon illustrates how grassroots movements and informal knowledge networks can significantly influence technological globalisation, technology adoption, and knowledge acquisition. The access to global networks, in turn, empowers them to contribute to the socioeconomic development of their regions, showcasing the pivotal role of uncodified knowledge circulations in global technological dynamics. In conclusion, while state-recognised forms of knowledge circulation have been extensively studied, the informal, uncodified practices facilitated by labour migration play an equally important role. These practices enable marginalised communities from Global South countries to actively participate in the globalisation of technology, driving sociotechnical transformations at the grassroots level.

ENDNOTES

1 The term Pasmanda, meaning 'those who have been left behind,' was first introduced by Ali Anwar (2023) to describe the socially and economically marginalised groups within the Muslim community in India. In his book *Masawat ki Jung: The Battle for Equality* (originally written in Hindi in 2001), Anwar explores the socio-political struggles of Pasmanda Muslims, providing a historical context and examining how these communities have been overlooked in both societal and political spheres. He delves into the caste dynamics within Indian Islam, challenging the perception of a monolithic Muslim identity and highlighting the internal discriminations faced by the Pasmanda. The book also discusses contemporary political strategies and potential alliances with other marginalised groups to promote social justice and equality in India.

2 The Kafeel, or the Kafala, system, also known as the sponsorship system, is a system of migrant labour governance that is predominantly used in the countries of the Gulf Cooperation Council (GCC), including Saudi Arabia, Qatar, Kuwait, Bahrain, the UAE, and Oman. The term 'kafala' translates as 'sponsorship' in Arabic. Under the Kafala system, migrant workers are tied to their employers, who act as their sponsors. This system regulates the entry, stay, and employment of migrant workers in the host country. The employer, or sponsor, known as 'Kafeel', holds significant control and authority over the migrant worker throughout their employment contract. But the boundaries between 'Kafeel' and 'Kafala' are often blurred, as migrants recognise the system (Kafala) through the man (Kafeel) who gives sponsorship.

3 For detailed guidelines regarding the Emigration Check Required (ECR) passport category, see Embassy of India, Riyadh, Saudi Arabia, "FAQs on ECR & non-ECR

(ECNR)," accessed March 18, 2024, https://www.eoiriyadh.gov.in/page/faqs-on-ecr-and-non-ecr-ecnr/.

4 For information about the eMigrate system, see Ministry of External Affairs, Government of India, "eMigrate," accessed March 18, 2024, https://emigrate.gov.in/.

5 Aadhaar: A 12-digit unique identification number issued by the Unique Identification Authority of India (UIDAI) to residents of India, serving as a proof of identity and address.

6 The platformisation of governance refers to the application of digital platforms and technologies within the framework of governance processes and public administration. It involves the use of digital platforms to enhance and transform how governments interact with citizens, deliver public services, and engage in decision-making processes (De Kloet et al. 2019; Masiero 2017).

REFERENCES

Aker, J. C., and I. M. Mbiti, 'Mobile Phones and Economic Development in Africa', *Journal of Economic Perspectives*, 24.3 (2010): 207–32.

Amirtham, N. S., & Kumar, A. Gender parity in STEM higher education in India: a trend analysis. *International Journal of Science Education, 43*(12) (2021): 1950–1968. Semantic Scholar+2https://niscpr.res.in+2ResearchGate+2

Anderson, W., and V. Adams, 'Pramoedya's Chickens: Postcolonial Studies of Technoscience', in E. J. Hackett, O. Amsterdamska, M. Lynch, and J. Wajcman, eds., *The Handbook of Science and Technology Studies* (3rd edn; MIT Press, 2008), pp. 181–204.

Antonio, A., & Tuffley, D. The Gender Digital Divide in Developing Countries. *Future Internet, 6*(4) (2014): 673–687.

Anwar, A., *Masawat Ki Jung: The Battle for Equality* (Forward Press, 2023) https://www.forwardpress.in/2023/03/masawat-ki-jung-the-battle-for-equality/.

Azhar, M., 'Indian Migrant Workers in GCC Countries', *Diaspora Studies*, 9.2 (2016): 100–111.

Beaudevin, C., and L. Pordié, 'Diversion and Globalization in Biomedical Technologies', *Medical Anthropology*, 35.1 (2016): 1–4 https://doi.org/10.1080/01459740.2015.1090436.

Behrends, A., S.-J. Park, and R. Rottenburg, *Travelling Models in African Conflict Management: Translating Technologies of Social Ordering*, 13 vols (Leiden: Brill, 2014).

Bélanger, D., and M. Rahman, 'Migrating Against All the Odds: International Labour Migration of Bangladeshi Women', *Current Sociology*, 61.3 (2013): 356–73.

Bhandari, R., 'Global Student and Talent Flows: Reexamining the Brain Drain Equation', *International Higher Education*, 99 (2019): 6–7.

Bhaskar, T., *Background Paper on Remittances from the GCC to India: Trends, Challenges and Way Forward*. Ministry of External Affairs, Government of India (2013).

Bhatty, Z., 'Social Stratification Among Muslims in India', in M. N. Srinivas, ed., *Caste: Its Twentieth Century Avatar* (Penguin Books, 1996), pp. 244–62.

Bijker, W. E., and J. Law, eds., *Shaping Technology/Building Society: Studies in Sociotechnical Change* (reissue edn; MIT Press, 1994).

Bowker, G. C., and S. L. Star, *Sorting Things Out: Classification and Its Consequences* (MIT Press, 2000).

Breman, J., *Capitalism, Inequality and Labour in India* (Cambridge University Press, 2019) https://doi.org/10.1017/9781108687485.

Chandramalla, C. B., 'Trends in Indian Labour Migration to Gulf Countries from 1990 to 2019: An Assessment of Literature', *Journal of Production, Operations Management and Economics*, 2.1 (2022): 1–9.

Chorev, N., and A. C. Ball, 'The Knowledge-Based Economy and the Global South', *Annual Review of Sociology*, 48 (2022): 171–91.

De Haas, H., *The Determinants of International Migration* (International Migration Institute, 2011).

De Kloet, J., T. Poell, Z. Guohua, and C. Yiu Fai, 'The Platformization of Chinese Society: Infrastructure, Governance, and Practice', *Chinese Journal of Communication*, 12.3 (2019): 249–56.

Dhar, B., and R. B. Bhagat, 'Return Migration in India: Internal and International Dimensions', *Migration and Development*, 10.1 (2021): 107–21.

Di Maria, C., and P. Stryszowski, 'Migration, Human Capital Accumulation and Economic Development', *Journal of Development Economics*, 90.2 (2009): 306–13.

Dumoulin Kervran, D., M. Kleiche-Dray, and M. Quet, 'Going South. How STS Could Think Science In and With the South?', *Tapuya: Latin American Science, Technology and Society*, 1.1 (2018): 280–305 https://doi.org/10.1080/25729 861.2018.1550186.

Fernandez, B., 'Essential yet invisible: migrant domestic workers in the GCC' Technical Report, Migration Policy Centre, European University Institute (2014) https://hdl.handle.net/1814/32148

Garratt, U., 'The Migration of Domestic Workers from the Philippines to the UAE and the Kafala System', *Academia Letters*, Article 1033 (2021).

Government of India. 'Social, Economic and Educational Status of the Muslim Community of India'. (2006) https://www.minorityaffairs.gov.in/WriteReadData/RTF1984/7830578798.pdf

Guérin, I. Juggling with debt, social ties, and values: The everyday use of microcredit in rural South India. *Current Anthropology*, 55(S9) (2014): S40–S50. https://doi.org/10.1086/675929

Gupta, N., & Sharma, A. K. Gender inequality in the work environment at institutes of higher learning in science and technology in India. *Work, employment and society*, 17(4) (2003): 597-616.

ILO, *India Labour Migration Update 2018* (2018) https://www.ilo.org/newdelhi.

Jena, P. K. Impact of pandemic COVID-19 on education in India. *International journal of current research (IJCR)*, 12 (7) (2020): 12582-12586

Jodhka, S. S., 'Caste in the Periphery', *Seminar*, New Delhi (2001): 41–46.

Kamat, A., and M. Crabapple, 'The Men in the Middle', *Dissent*, 62.2 (2015): 65–72.

Kathiravelu, L., 'Social Networks in Dubai: Informal Solidarities in an Uncaring State', *Journal of Intercultural Studies*, 33.1 (2012): 103–19.

Khadria, B., 'India: Skilled Migration to Developed Countries, Labour Migration to the Gulf', *Migración y Desarrollo*, 7 (2006): 4–37.

Khan, A., and H. Harroff-Tavel, 'Reforming the Kafala: Challenges and Opportunities in Moving Forward', *Asian and Pacific Migration Journal*, 20.3–4 (2011): 293–313.

Levitt, P., 'Social Remittances: Migration-Driven Local-Level Forms of Cultural Diffusion', *International Migration Review*, 32.4 (1998): 926–48 https://doi.org/10.1177/019791839803200404.

Lo, L., W. Li, and W. Yu, 'Highly-Skilled Migration from China and India to Canada and the United States', *International Migration*, 57.3 (2019): 317–33.

Madianou, M., & Miller, D. *Migration and new media: Transnational families and polymedia.* (Routledge, 2013)

Malaeb, H. N., 'The Kafala System and Human Rights: Time for a Decision', *Arab Law Quarterly*, 29 (2015): 307.

Martin, A. K. Between surveillance and recognition: Rethinking digital identity in aid. *Big Data & Society*, 8(1) (2021): 1–12. https://doi.org/10.1177/20539517211006744

Masiero, S., 'Digital Governance and the Reconstruction of the Indian Anti-Poverty System', *Oxford Development Studies*, 45.4 (2017): 393–408.

Masiero, S., 'Subaltern Studies: Advancing Critical Theory in ICT4D'. *Proceedings of the European Conference on Information Studies* (2018). https://aisel.aisnet.org/ecis2018_rp/162

Mavhunga, C. C., *Transient Workspaces: Technologies of Everyday Innovation in Zimbabwe* (MIT Press, 2014).

Mavhunga, C., ed., *What Do Science, Technology, and Innovation Mean from Africa?* (MIT Press, 2017) https://doi.org/10.26530/oapen_631166.

Nakamura, L., *Digitizing Race: Visual Cultures of the Internet*, 23 vols (University of Minnesota Press, 2007).

Parreñas, R. S., and R. Silvey, 'The Governance of the Kafala System and the Punitive Control of Migrant Domestic Workers', *Population, Space and Place*, 27.5 (2021): e2487.

Pollozek, S., J.-H. Passoth, and H. Dijstelbloem, 'Data Circulation and the (Re) Configuration of European Migration and Border Control: A Praxeographic Inquiry into the Information Infrastructure of the Frontex Joint Operation Poseidon', *European Journal of Migration and Law*, 23 (2021): 188–212.

Rajan, S. I., and P. Kumar, 'Historical Overview of International Migration', in S. I. Rajan, ed *Governance and Labour Migration* (Routledge India, 2020), pp. 1–29.

Rao, U., & Nair, V. G. Aadhaar: Governing with biometrics. *South Asia: Journal of South Asian Studies*, 42(3) (2019): 469–481. https://doi.org/10.1080/0085 6401.2019.1595343

Reserve Bank of India. *Report of the Committee on Deepening of Digital Payments.* (2018). Retrieved March 18, 2024, from https://www.rbi.org.in/Scripts/PublicationReportDetails.aspx?UrlPage=&ID=892

Rosenberg, K., B. K. Karki, and R. Kurian, 'Financial Literacy is Key for Optimizing the Development Potential of Remittances', *2017 EADI–Nordic Conference* (2017).

Rottenburg, R., N. Schräpel, and V. Duclos, 'Relocating Science and Technology: Global Knowledge, Traveling Technologies and Postcolonialism', *Perspectives on Science and Technology Studies in the Global South* (Max Planck Institute for Social Anthropology, 2012) http://www.hsozkult.de/event/id/event-68318.

Saxenian, A., 'From Brain Drain to Brain Circulation: Transnational Communities and Regional Upgrading in India and China', *Studies in Comparative International Development*, 40 (2005): 35–61.

Saxenian, A., and J.-Y. Hsu, 'The Silicon Valley–Hsinchu Connection: Technical Communities and Industrial Upgrading', *Industrial and Corporate Change*, 10.4 (2001): 893–920.

Shah, N. M., and I. Menon, 'Chain Migration through the Social Network: Experience of Labour Migrants in Kuwait', *International Migration*, 37.2 (1999): 361–82.

Shevinsky, E., *Lean Out: The Struggle for Gender Equality in Tech and Start-Up Culture* (OR Books, 2015).

Singh, R. Give me a database and I will raise the nation-state. *South Asia: Journal of South Asian Studies*, 42(3) (2019): 501–518. https://doi.org/10.1080/0085 6401.2019.1602810

Smith, L. T., *Decolonizing Methodologies: Research and Indigenous Peoples* (Zed Books, 2021).

Srinivas, M. N., 'An Obituary on Caste as a System', *Economic and Political Weekly* (1–7 Feb. 2003): 455–59.

Srivastava, R., 'Labour Migration, Vulnerability, and Development Policy: The Pandemic as Inflexion Point?', *The Indian Journal of Labour Economics*, 63 (2021): 163–85 https://doi.org/10.1007/s41027-020-00301-x.

Sultana, H., and A. Fatima, 'Factors Influencing Migration of Female Workers: A Case of Bangladesh', *IZA Journal of Development and Migration*, 7 (2017): 1–17.

Tsai, K. S., 'Social Remittances of Keralans in Neoliberal Circulation', in S. Rajan, ed., *India Migration Report 2020* (Abingdon: Routledge India, 2020), pp. 228–52.

Twagira, L. A., 'Africanizing the History of Technology Introduction', *Technology and Culture*, 61.2 (2020): S1–S19.

Upadhya, C., *Reengineering India: Work, Capital, and Class in an Offshore Economy* (Oxford: Oxford University Press, 2016).

Vertovec, S., 'Migrant Transnationalism and Modes of Transformation', *International Migration Review*, 38.3 (2004): 970–1001 https://doi.org/10.1111/j.1747-7379.2004.tb00226.x.

Vickstrom, E., 'Pathways into Irregular Status among Senegalese Migrants in Europe', *International Migration Review*, 48.4 (2014): 1062–99 https://doi.org/10.1111/imre.12154.

Von Schnitzler, A., 'Traveling Technologies: Infrastructure, Ethical Regimes, and the Materiality of Politics in South Africa', *Cultural Anthropology*, 28.4 (2013): 670–93 https://doi.org/10.1111/cuan.12032.

Williams, A. M., 'International Labour Migration and Tacit Knowledge Transactions: A Multi-Level Perspective', *Global Networks*, 7.1 (2007): 29–50.

Winner, L. 'Do artifacts have politics?' Daedalus 109 (1) (1980):121--136. https://www.jstor.org/stable/20024652

Wise, R. D., 'Reframing the Migration and Development Question in the Twenty-First Century', in A. Bloch and C. Levy, eds., *Routledge Handbook of Immigration and Refugee Studies* (Routledge, 2022), pp. 319–28.

Zachariah, K. C., & Rajan, S. I. *Dynamics of Emigration and Remittances in Kerala: Results from the Kerala Migration Survey 2014* (Working Paper No. 463: (2016)). Centre for Development Studies. Retrieved from https://cds.edu/wp-content/uploads/WP463.pdf

Zweig, D., K. S. Tsai, and A. D. Singh, 'Reverse Entrepreneurial Migration in China and India: The Role of the State', *World Development*, 138 (2021): 105192.

NAVIGATING INTERNATIONAL INEQUITIES

4

CALIBRATING THE GLOBAL: HOW ARE GHANAIAN SCIENTISTS SHIFTING AFRICA'S POSITION IN GLOBAL ATMOSPHERIC SCIENCE?

Jessica Pourraz and Allison Felix Hughes

ACCORDING TO THE WORLD HEALTH ORGANISATION (WHO), AIR POLLUTION – ambient and indoor – is the leading environmental risk to health worldwide. It accounts for 7 million deaths per year and most severely affects populations in the Global South, where nearly 92% of total air pollution–related deaths occur. It also reflects significant social inequalities, as the poorest populations are the most affected (Landrigan et al. 2017). Prospective studies show, for instance, that by 2050, the African population will almost double, and half of its inhabitants will live in cities where they will more likely be exposed to air pollution (Katoto et al. 2019). In African cities, the main contributors to air pollution are vehicle fuel combustion; dust from unpaved roads, brickwork, and construction sites; the combustion of biomass and household waste; and industry.

In 2018, the WHO estimated that in the capital city of Ghana, Accra, 28,000 premature deaths had occurred due to air pollution. Despite the scale and severity of air pollution in Ghana, institutional monitoring systems remain

weak, and environmental regulations are poorly enforced. To measure air pollution in the metropolis of Accra, the experts at the Environmental Protection
Agency (EPA-Ghana) only have a few outdated mobile devices, which they use
along the main roads, and two fixed reference-grade monitors located on the
University of Ghana campus and at an elementary school in downtown Accra.
The monitoring improvement is made all the more difficult as Global North
actors set international standards and rely upon expensive reference-grade
monitors. These monitors are high-precision air-quality control instruments
that meet specific standards and criteria set by internationally recognised
regulatory agencies and scientific organisations. They undergo rigorous testing, calibration, and quality assurance procedures to ensure their accuracy and
traceability in compliance with recognised standards. These monitors often
involve advanced sensor technology and may incorporate multiple sensors to
measure various air pollutants simultaneously. Reference-grade monitors are
heavy machinery installed in a fixed location (they are not portable), and they
are very costly (US$100,000 on average). In the field of air pollution research,
the international scientific community and academic journals consider highly
technical and expensive measurement instruments such as fixed reference-
grade monitors – not affordable to most scientists in the Global South – as the
'gold standard'.[1] 'Not only does this virtually create a Global North monopoly
on publication, but it also wrongly positions research carried out with such
technology as "novel" as if the new research was shedding light on an unknown
problem' (Negi and Ranjan 2020: 2).

In turn, the lack of resources to purchase, operate and maintain the traditional fixed reference-grade monitors largely explains the dearth of regular
and systematic air quality monitoring in countries like Ghana – as well as
in most of the other Sub-Saharan African (SSA) countries (Giordano et al.
2021; Hagan and Kroll 2020; Morawska et al. 2018). This issue directly affects
Ghanaian scientists working on air pollution, but does not mean that air pollution research remains out of reach of Ghanaian scientific institutions: studies
of air pollution in countries like Ghana (Arku et al. 2008; Dionisio et al. 2010)
and Kenya (deSouza 2020) go back several decades and involve considerable
work carried out by local researchers, who adapted their methods to less costly

material. How do scientists from these countries manage to build expertise in air pollution?

One key tool used by Ghanaian researchers has been low-cost sensors (LCS), which have become very popular among local academic actors and even among EPA-Ghana experts, because they make it possible to produce air quality data at a much lower cost. LCS are affordable miniature low-tech sensors designed to produce real-time continuous data on particulate matter (PM) – a mixture of solid particles, liquid droplets, and gases – that can be shared through digital platforms. These devices illustrate the emergence of a growing but highly differentiated global market for air monitoring. The air quality monitoring market worldwide accounted for US\$5.08 billion in 2024 and is expected to reach US\$6.73 billion by 2029, with a growth rate of almost 6% between 2024 and 2029[2]. This market is highly segmented by product type, end user, and geography. It is also deeply hierarchised through norms and standards. It includes both expensive reference-grade monitors and LCS, the latter being mostly dedicated to low-income countries, such as in the Global South, where they form alternative or even complementary networks to the institutional monitoring system.

To a certain extent, the LCS market constitutes a limited and temporary solution for Ghanaian experts, as it implies both a prolonged dependence on gold standards and certain forms of exploitation by the LCS-providing companies. Indeed, some experts question the validity of LCS measurements and point to the fact that, unlike reference-grade monitors, LCS are not subject to international standards and certification procedures. Therefore, there is a need to benchmark LCS against fixed reference-grade monitors (Parasie and Dedieu 2019; Sahu et al. 2020). This work of benchmarking, also called calibration, is 'defined as the process by which outcomes from LCS are compared against reference monitors and adjusted after comparison' (Pritchard et al. 2018: 6). In order to make the devices accurate and ensure the validity of the data produced, Ghanaian scientists carry out this calibration by themselves; this calibration work is, in turn, appropriated and valued on the market by LCS companies that sell their devices to other Global South countries. Ghanaian scientists, therefore, not only depend on a scientifically dismissed market, defined by lower standards than the 'gold standard' they cannot afford in a knowledge economy dominated by

Global North actors, but the calibration work they achieve in this market is not officially recognised nor acknowledged, nor financially compensated despite the added value it brings to the devices for African markets. But at the same time, the LCS market operates as an emancipatory location for Ghanaian scientists. The use and calibration required by LCS play an important part in shaping expertise in the Global South. They help Ghanaian scientists shift Africa's position within the global atmospheric science community by producing scientific knowledge and publishing it in international journals.

By analysing the assemblages formed by academics from the Global North and African experts, international donors, and private actors to produce scientific knowledge, this chapter explores the complexity of this situation. It shows how Ghanaian scientists are coping with international inequities in air knowledge production by inserting themselves into global technological markets and by appropriating incoming technologies, instruments, and tools from the Global North to develop legitimate expertise and create new ways of conducting atmospheric science. With these issues in mind, we (1) describe the international academic partnerships that have been established over the last fifteen years in Ghana to allow their scientists to conduct research; (2) discuss calibration in practice by examining the invisible work of the Air Quality Research Laboratory of the Department of Physics at the University of Ghana, where environmental and atmospheric physicists carry out all the calibration work; and (3) conclude by analysing how this free and invisible labour participates in shaping Ghanaian scientific expertise and contributes to legitimising Ghanaian scientists in global atmospheric science.

This contribution is based on an ethnographic work carried out at the Air Quality Research Laboratory of the Department of Physics at the University of Ghana (UG), at the laboratory of the Environmental Quality Department of the EPA-Ghana, and during their air quality–monitoring tour of the Accra metropolis between August 2021 and April 2023. These observations were completed by semi-structured interviews conducted in Accra (N=35) with atmospheric physicists and academics from the University of Ghana, EPA experts, members of NGOs, representatives from startups, officials from the Greater Accra Metropolitan Area, and officials from the Ministry of Environment. It was completed by the analysis of grey literature, such as EPA-Ghana activity reports, press articles, and social media posts.[3]

'AFRICA–GLOBAL NORTH' RESEARCH COLLABORATIONS ON AIR POLLUTION IN ACCRA

Air pollution stems from a combination of several emission sources, which specific meteorological and geographical conditions may amplify. Particulate matter (PM) (PM2.5 and PM10) is the most commonly used measure of air pollution. PM10 includes particles with a diameter of less than ten micrometres (μm), while PM2.5, or fine particles, are those smaller than 2.5 μm. Epidemiologists have identified the adverse effects of fine particles on human health due to their ability to penetrate deep into the lungs, affecting the respiratory system on a deeper level (Moizard-Lanvin 2021). PM data are, therefore, the most commonly produced to objectify air pollution and the easiest data to collect, as they do not require any laboratory chemicals or organic analysis.

West African cities such as Accra have PM levels well above WHO thresholds. The WHO recommends that PM2.5 should not exceed ten micrograms per cubic metre of air (μg/m3) and that PM10 should not exceed 20 μg/m3 (World Health Organisation 2021). However, average levels in Accra are still two to four times higher than the WHO guideline value (Alli et al. 2021).

The issue of air pollution is not new in Ghana. The first Environmental Protection Act – EPA Act 490 – was passed in 1994. In 1997, the Environmental Quality Department of the EPA-Ghana, under the remit of the Ministry of the Environment, began monitoring air quality as part of a programme funded by the World Bank, which ended in 2001. Several air quality indicators were measured on an ad hoc basis in the western region of Ghana, which was polluted by mining activities, and in some districts of the main Ghanaian cities, including Accra, Kumasi, Takoradi, and Tema. In 2004, the EPA-Ghana experts received support from the US Agency for International Development (USAID) and the US Environmental Protection Agency (US-EPA) to deploy an institutional system for monitoring PM in residential, commercial, and industrial areas of Accra and along the main roads of the Greater Accra Metropolitan Area (Pourraz 2024).

Around this time, in 2006, the now head of the Quality Research Laboratory of the UG Department of Physics began conducting research on air pollution as part of his PhD on the characterisation and source apportionment of air-borne particulate matter in certain urban neighbourhoods of Accra.[4] He had the opportunity to work under the supervision of Professor Majid Ezzati at the Harvard School of Public Health in Boston. He thus benefited from the material and instrumental resources of this prestigious university. Prior to that, academic research in this area in Ghana was non-existent.

This work has led to a number of scientific collaborations, enabling Ghanaian academics from the Department of Physics and the Department of Geography at UG to conduct studies in Accra with the support of Professor Ezzati of Harvard University. Thus, for about a month in 2006, these academics collected continuous data on PM2.5, PM10, and concentrations of sulphur dioxide (SO_2) and nitrogen dioxide (NO_2) – which are indicative of fine variations in vehicle pollution – in two poor neighbourhoods of Accra: James Town and Nima. The first results of this pioneering research highlighted vehicles' contribution to air pollution. The academics' next study, carried out in 2007, further investigated the trends in these same neighbourhoods, with stationary and mobile data collected over three months. In September 2007, two additional neighbourhoods – Asylum Down (middle class) and East Legon (upper class) – were added to the study and monitored for a year. The results showed that biomass particles, road dust, and vehicle emissions were the main contributors to PM (Arku et al. 2008; Dionisio et al. 2010a; Dionisio et al. 2010b; Rooney et al. 2012; Zhou et al. 2013). For all these studies, the monitors used to measure PM were considered not as LCS but as reference sensors, even though they were not fixed reference-grade monitors.

These academic collaborations initiated in 2006 have continued as part of the global partnership Pathways to Equitable Healthy Cities, coordinated since 2018 by Professor Ezzati, who is now working at the School of Public Health of Imperial College London. Within the framework of this global partnership, several students from the UG have been recruited to conduct their MPhil research, or even as research assistants, to collect weekly data (PM, weather variables, pictures) and maintain the database. These collaborations have been reinforced by the cooperation of US-EPA experts and scholars from Columbia University, York University, and the Kigali Collaborative Research Centre, through which Ghanaian academics were able to obtain a significant number of LCS. Fifteen years of this 'Africa–Global North' scientific collaboration have enabled Ghanaian scientists to make their mark on the global academic landscape by performing measurements of the spatial, socioeconomic, and temporal patterns of ambient and indoor air pollution in Accra, and by differentiating pollutants and their sources, using mainly LCS.

In January 2019, the EPA-Ghana experts, with the help of Ghanaian academics and financial support from the World Bank, finally installed the first two fixed reference-grade monitors in Accra, including one at the UG Department of Physics. Fixed reference-grade monitors play an instrumental role, as they are designed to provide accurate and reliable measurements of various air pollutants. They are also used as benchmark instruments for comparing or calibrating other air quality–monitoring devices such as LCS. By that point, EPA-Ghana experts had already deployed ten LCS in Accra, including four received from the US-EPA to measure PM on a continuous basis. Given the lack of sufficient financial resources, these LCS allowed for the expansion of the institutional monitoring network and the characterisation of the spatial variability of PM levels in different areas of Accra.

Because Ghana is located close to the equator, with extreme temperatures and other particular weather parameters, the Air Quality Research Laboratory of the UG Department of Physics is striving to become a centre of excellence for LCS testing. Its unique technological setup, including a fixed reference-grade monitor, should encourage more academic collaborators from abroad to have their LCS tested by the UG Department of Physics and make them suitable for use elsewhere in Africa. This whole process is part of a wider project conducted by the UG Department of Physics to replicate the Air Quality Sensor Performance Evaluation Center (AQ-SPEC) programme established by the regulatory agency responsible for improving air quality in California and funded by the US-EPA. As part of an effort to inform the general public, the AQ-SPEC programme aims to conduct a thorough characterisation of currently available LCS under ambient (field) and controlled (laboratory) conditions.[5] Ghanaian environmental and atmospheric physicists are therefore seeking not only to create a hub to test LCS and publish the results of their evaluations online but also to train more academics and experts from Africa to use and calibrate LCS.

In this first section, we have shown how, over the last 15 years, Africa–Global North research and institutional collaborations surrounding air pollution have enabled the training of Ghanaian scientists, the development of local capacities, and the acquisition of instruments. Transnational technological, financial, and human flows to Ghana, mainly from the United States and later from the United Kingdom, highlight the dominance of the anglosphere and the Global North in international research on air pollution, which is also reflected in the financial support provided by the World Bank and USAID and technical assistance from the US-EPA to the EPA-Ghana experts. Having said that, by appropriating financial resources, instruments, and knowledge from the Global North, Ghanaian scientists are trying to counterbalance this dominance and become key players in atmospheric science in Africa, for instance by replicating the AQ-SPEC.

In the following sections, we show how Ghanaian scientists, by adapting LCS to local settings via calibration work, are instrumental in producing knowledge surrounding air quality and are contributing to shifting Africa's position in global science.

RECONFIGURING AIR QUALITY MONITORING IN THE GLOBAL SOUTH WITH LCS

The need for more air quality data for African cities has severely hampered efforts to characterise and understand the patterns of air pollution concentrations and to promote policy initiatives to control and regulate such pollution. Until 2010, only actors such as governments and research organisations that could afford very expensive instruments and had qualified personnel could routinely perform PM measurements similar to those taken in the Global North.

Over the last decade, however, air quality monitoring has seen radical changes, driven by the emergence of LCS, which provide a way to bridge the gaps in official air quality monitoring. LCS were originally intended to democratise environmental monitoring practices, such as air quality sensing, that might ordinarily be the preserve of expert scientists. 'Citizen science', which refers to this alternative production of knowledge around toxic exposure by a wide range of actors without resources – such as citizen groups, activists, and vulnerable populations exposed to toxic sites – has been studied extensively by social science academics in the United States, in the context of petrochemical industrial areas in Louisiana and Pennsylvania (Allen 2018; Gabrys et al. 2016; Ottinger 2010; Pritchard and Gabrys 2016). These works have demonstrated that most of the data produced by LCS are intended for information sharing, awareness raising, and lobbying the authorities, and not yet for any legal purposes.

In resource-constrained countries around the world, such as those in Sub-Saharan Africa, the use of LCS for air quality monitoring offers significant opportunities to expand air quality data acquisition. This has led to a reconfiguration of the instruments and actors able to perform such monitoring – with the latter now including academics in the Global South – and of 'the temporal and spatial scales on which we think about air quality in general' (Giordano et al. 2021: 158). LCS can provide measurements on the level of a street or a neighbourhood, that is, at a scale that previously eluded monitoring institutions. These sensors also enable continuous monitoring in real-time, which can show pollution peaks and not just average pollution levels, as is often the case with institutional air monitoring (Ottinger 2010). 'These techniques also are reported to neglect the prospect of being able to correlate the variations in short-term intra-day atmospheric parameters' (Sahu et al. 2020: 1347). In other words, low-cost technologies have allowed for the atmospheric sciences to become more accurate and precise.

In Accra, academics and EPA experts are deploying a wide range of LCS, all portable direct-reading PM2.5 monitors, such as the Chinese ZeFan continuous monitor and the US monitors Clarity Node, PurpleAir, and QuantAQ Modulair. These LCS are all based on light-scattering techniques to measure mostly PM2.5. As Figure 4.1 shows, 'particles flow through the measurement cavity, the light intensity of the infrared/red light reaching the phototransistor is modulated by the presence of particles in the light path' (Giordano et al. 2021: 3). Put differently, the laser detects the particles and counts them (as can be seen in Figure 4.1, representing the resulting output for the photodiode signal in a low-cost PM sensor with particles absent (above) and present (below) in the light path).

These LCS are also integrated into digital infrastructures (computer servers, mobile applications, websites) that make the circulation, processing, and formatting of data possible (Parasie and Dedieu 2019). LCS require infrastructure comprising an electrical network, Wi-Fi, and data-management centres. Start-up companies are stepping up to produce affordable, easy-to-use, and portable wireless PM sensors to monitor air pollution (Sahu et al. 2020). These developments are shaping a new market for air data.

Clarity Node and PurpleAir are the LCS most commonly used by academics working in Accra to measure PM continuously and in real-time. Both companies sell their LCS at the same price and under the same conditions, irrespective of who the customer is and where they are located in the Global North or Global South. Beyond this common trait and the fact that they are both American, the two companies have differing market models and strategies.

PurpleAir is considered to be the pioneer in the field of LCS. Its device requires Wi-Fi access to upload the data to the cloud and a power source. This can be highly challenging in Africa, where power cuts and connectivity issues are common. Nevertheless, there is the option to store data on an SD card. As

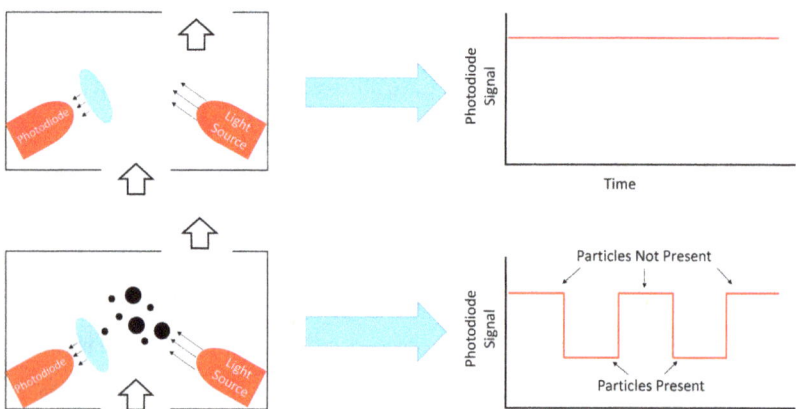

FIG. 4.1 Light-scattering techniques to measure particulate matter (PM) (Giordano et al. 2021).

this is an open data system, there is no need for a license to use the LCS. The cost of the device is around US$250, which is very cheap in comparison to the price of the reference-grade monitors, but still represents a significant amount of money for citizens of the Global South and at the margins. PurpleAir sells sensors to a wide range of actors, such as private citizens, academics, and NGOs working to develop citizen science. On its website, the company claims that it 'makes sensors that empower community scientists who collect hyper-local air quality data and share it with the public'.[6]

Clarity was created in 2014 in collaboration with UC Berkeley.[7] It focuses on PM2.5 and NO_2, and its devices are manufactured in Taiwan for economically advantageous reasons related to contract manufacturing conditions. Clarity Node devices are all solar-battery powered and have a cellular connection, which makes them significantly easier to use in places where it can be challenging to maintain infrastructure, both in developed economies and in lower- to middle-income countries.

One of the key differences, compared to PurpleAir, is that Clarity does not sell sensors to private individuals and, therefore, does not support individual citizen scientists. Clarity works directly with government agencies, researchers, and organised community groups, selling them its devices. This is because Clarity's experts want direct access to the reference data produced by regulatory authorities, for instance, with the two fixed reference-grade monitors belonging to the EPA-Ghana. Accumulating data from these monitors would enable Clarity technicians to calibrate their low-cost devices and to ensure that their data are accurate or, if necessary, to apply correction factors. This would give them total control over the system and allow the company to create even more value.

According to Clarity's official statement, citizen science–type models cannot achieve this level of quality assurance and quality control around data correction and calibration. This is partly why Clarity likes to partner with government agencies. The company can then design and deploy LCS networks in connection with specific sources, policies, and interventions to objectively support certain operational decisions. Clarity considers that this is where LCS are most useful: used together with regulatory monitoring as an extension or a supplement.

This approach explains why, contrary to PurpleAir, the Clarity Node LCS are not the property of their users. Customers have to pay an annual service subscription fee (licence fee) of US$1,200 – a relatively high cost for academics in the Global South – which includes hardware, hardware replacements, software, cloud services, and full-time project support. This license has to be renewed every year, which can be a financial obstacle to its use by academics, among other actors. The data are the property of the clients, although they can be made available to everyone, including the company, as open access data on a non-profit platform, Open AQ (https://openaq.org/#/), and on Clarity Open Map, a simplified platform. The idea, with these platforms, is to pool all the data and standardise them, with a view to supporting global or regional data sharing.

Clarity acts as the custodian of the data that are kept on its servers and does not monetise the data in the sense of selling them. The company supports Open Data, which it feels constitutes a powerful tool for improving air quality management and fostering community engagement. However, air quality data and the data collected on how customers use the sensors help the company improve its LCS operation. The company can thus monetise these improvements by increasing their licence fees and thereby creating market value. For example, Clarity has fitted its LCS with mini-solar panels to ensure their energy self-sufficiency, as well as an SD card slot to alleviate internet connectivity problems.

As part of the Pathways to Equitable Healthy Cities partnership, Ghanaian and US academics have also deployed a ZeFan device, a Chinese low-cost real-time continuous monitor that measures PM2.5 concentrations at one-minute intervals, which is extremely basic (Alli et al. 2021). Minute-by-minute measurements of weather variables are also recorded by a portable weather meter that tracks several weather parameters, including temperature, relative humidity (RH), and the heat index, which can affect the accuracy of air pollution data. Results show that at all sites, PM2.5 peaks at dawn and dusk, coinciding with rush-hour commuting, and that nitrogen oxide (NO_x) levels, mainly from vehicles, are increasing.

Although the forms and presentations of these LCS vary, from the simplest device, such as ZeFan, to the most elaborate one – such as the Clarity Node with its solar panel – all contain the same components. The main one, and the

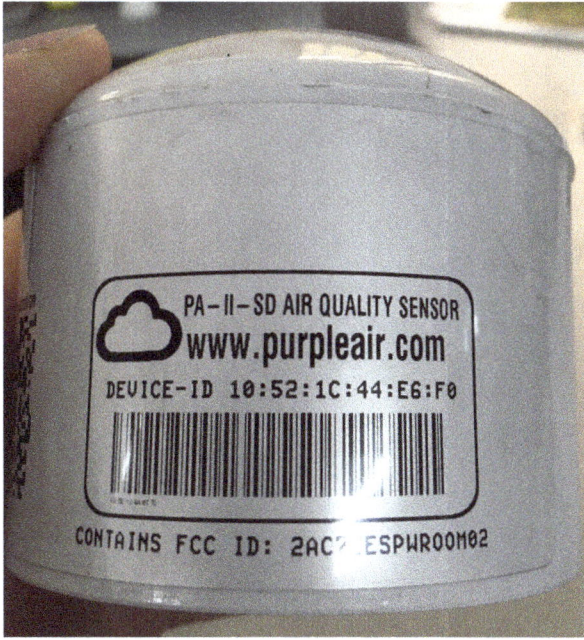

FIG. 4.2A PurpleAir sensor (Jessica Pourraz)

FIG. 4.2B Clarity Node-S sensor (permission from Clarity Node)

FIG. 4.2C ZeFan sensor 88 (Jessica Pourraz)

FIG. 4.2D Modulair sensor (Jessica Pourraz)

most important, is the Plantower sensor, which is made in China.[8] It consists of a small rectangular metal case containing a miniature fan, a printed circuit board, and a laser (as shown in Figure 4.3).[9] All companies producing LCS around the world, whether American or Indian, depend on China to procure this sensor.

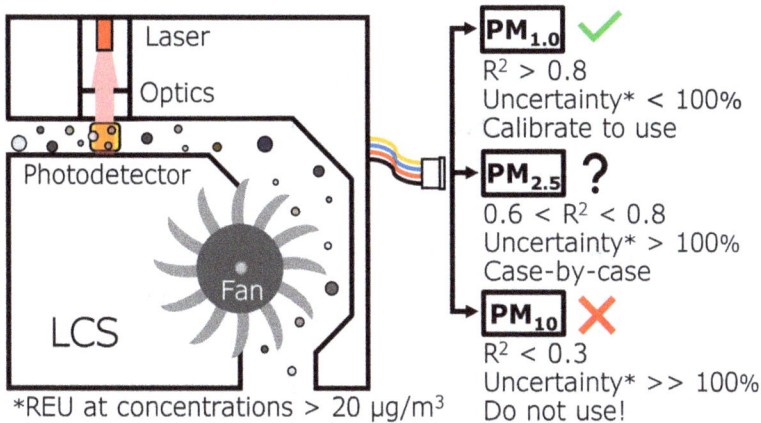

FIG. 4.3 Plantower sensor, which consists of a small rectangular metal case containing a miniature fan, a printed circuit board, and a laser (Rueda et al., *Environ Sci Technol Lett* 10 [2023])

The performance of these LCS in the ambient atmospheric environments in which they are being used, however, has yet to be thoroughly evaluated (Lewis et al. 2016). Some of the LCS are far from efficient, even with correction factors applied, and most of them deteriorate very quickly when exposed to humidity and very high temperatures. The performance and accuracy of LCS are significantly affected by particle-size distribution, which changes depending on the source of the pollution and by the variability of meteorological parameters like temperature and humidity, which can be extreme in Global South countries such as Ghana. Data from uncalibrated LCS are insufficient to inform policy decisions. Thus, in addition to testing sensors in situ in order to adjust them to the local environment (Pritchard et al. 2018), calibration work is necessary. In the last section, we present and analyse this calibration work in practice, and the role of Ghanaian scientists, by examining the Air Quality Research Laboratory of the UG Department of Physics.

FROM FREE AND INVISIBLE SCIENTIFIC LABOUR TO LEGITIMISED GHANAIAN SCIENCE

One of the authors of this paper carries out the LCS calibration process himself, as observed by the other author. Calibration is a fairly straightforward process, which involves positioning LCS alongside fixed reference-grade monitors for a certain amount of time – between three weeks and four months – in order to ensure that the data they measure are of high quality. The ultimate goal is to identify correction factors that can be applied to correct these LCS when they are over-recording or under-recording. It also helps to test their robustness in such extreme conditions.

The data collected from the LCS and reference-grade monitor are then downloaded onto the UG Physics Department computers. The computers are equipped with specific software used to analyse the data and convert them into a graph made up of curves. The curves showing the respective PM data from the LCS and the fixed reference-grade monitor are displayed on the screen and compared. Even if they do not overlap exactly, the curves should at least follow the same trend. The correction process is then carried out to reduce the discrepancy between the two curves (the blue and black in Figure 4.4b), often due to under- or overestimation by the LCS. Calibration allows for different types of knowledge to be held alongside each other.

In 2021, one of the authors of this paper carried out one of the first LCS intercomparison studies in Africa at the UG Department of Physics. For four months, he and his team collocated twenty LCS – two QuantAQ Modulair, two PurpleAir, and sixteen Clarity Node devices – with the fixed reference-grade monitor on the department's rooftop. The LCS were also collocated with a meteorological station measuring temperature, relative humidity, and wind speed and direction.

The collocation site on the university campus is an urban area with low-density housing, few trees, and relatively limited traffic flows. The nearest road is 500 metres away. There are no known large combustion facilities or other emission sources near the site. The objectives of the study were to assess the LCS performance in terms of precision and reliability, to evaluate how each

FIG. 4.4A Collocation at UG of LCS (located in the two posts in front of the picture) with fixed reference-grade monitors, which are located inside the white shelter at the back of the picture (Jessica Pourraz)

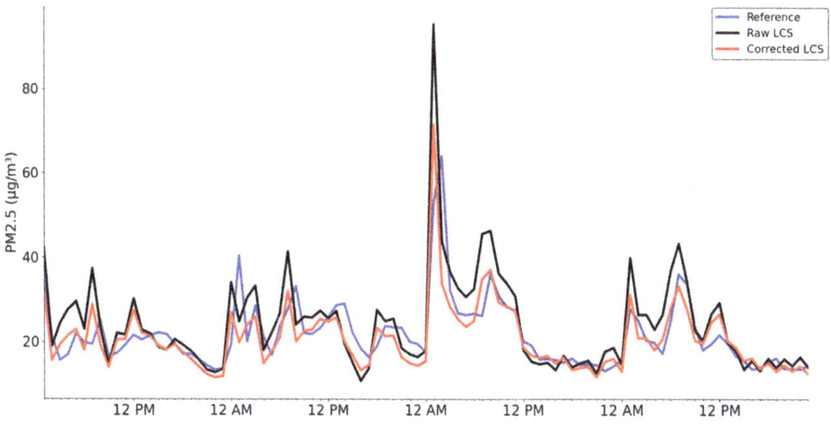

FIG. 4.4B Calibration process, Accra, Ghana, August 2021 (Allison Felix Hughes)

type of LCS correlated with the fixed reference-grade monitor, and to compare the use of four machine learning models to correct the LCS data. Machine learning is a branch of artificial intelligence (AI) that can learn from feedback (McFarlane et al. 2021). The Ghanaian scientists have used the data produced by the reference monitor and the LCS to develop the most accurate correction factors for calibrating the LCS data. Doing so, these LCS could be deployed at other locations, and the correction-factor model would correct the data to reflect the values that a reference monitor would have produced at that location.

Each LCS model requires its correction factor, which explains why academics developed four different correction factors for each brand of LCS. The four correction factor models that were developed for PurpleAir, Clarity, and QuantAQ Modulair were then applied to correct PM2.5 data collected using a network of LCS situated around Accra. Thus, if the LCS were overestimating or underestimating values, those biases would be removed by the correction factor equation developed by the academics at UG. In order to stabilise the data, two LCS remain permanently collocated with the fixed reference-grade monitor at UG. As pollution sources change with the seasons, the LCS regularly need to be re-corrected. If the LCS data are not corrected, or if the devices are not used or tuned properly, they can be very misleading, which can generate inaccurate data.

The private companies producing the LCS do not, however, financially compensate for all the calibration work described above and performed by Ghanaian scientists. The literature has already provided critical insights into the free labour exploited through do-it-yourself projects and the maker movement (Davies 2017). In our research, we found that this was an issue not only of free labour, but also of invisible labour (Bangham et al. 2022) by Global South scientists for the benefit of LCS manufacturers located in the Global North. Ghanaian academics carry out, free of charge, all the testing and validation work for LCS manufacturers, completely anonymously, and invisibly for the other customers who benefit from it:

> So we actually have full-time project partners working with the Clarity
> networks, working with the Ghana EPA, helping establish the correction
> factors, diagnosing technical challenges, replacing faulty devices, providing

siting guidance, etc. And it's very helpful in terms of providing at fixed costs for technical deployments with kind of reducing risks around network failures, connectivity issues. We try to take all these little costs and challenges associated with scaling air quality data and make it very predictable and easy to fund in a budget. So to speak. (VP, Business Development and Partnerships, Clarity Movement Co).

In other words, all the free and invisible labour provided, and the knowledge produced, by Ghanaian academics and scientists is leveraged by the private company to put together a package of services, which it will then sell to these same experts and scientists when they buy their annual service subscription (license) at a cost of US$1,200. Ghanaian scientists are, therefore, part of what some scholars have called 'polycentric innovation', a term which 'designates the global integration of specialised research and development capabilities across multiple regions to create novel solutions that no single region or company could have completely developed on its own' (Bhaduri 2016: 8). However, they don't appear as co-developers of innovation, let alone benefit from the economic spin-offs generated by the sale of LCS.

At the same time, since 2006, access to and the calibration of these new technologies has allowed Ghanaian researchers to produce and publish scientific data on air quality (Alli et al. 2021; Arku et al. 2008; Clark et al. 2020; Dionisio et al. 2010a; Dionisio et al. 2010b; Jack et al. 2015; McFarlane et al. 2021; Rooney et al. 2012; Zhou et al. 2013). Ghanaians indeed use insights from data produced by low-cost technologies from the Global North and China, which enable them to develop 'legitimate' expertise. By doing so, they exist – and do not perish (Harzing 2010) – in an academic landscape largely dominated by researchers from the Global North who can afford fixed reference-grade monitors for their research. However, this legitimacy is still from the perspective of Global North actors who define what legitimate data are and are not (Garrocq and Parasie 2022). The balance of power remains with Global North knowledge production, despite the active role of Ghanaian scientists, who are still dependent on the values and perspectives of the Global North. Indeed, 'since the academy evaluates knowledge based on Western standards of reliability and validity, non-Western paradigms will still have to be altered to fit the criteria of Western frameworks' (Thambinathan and Kinsella 2021: 2). Nevertheless, this allows for Africa's position in global atmospheric science to shift and has, for instance, led to the organisation of the first Air Sensors International Conference (ASIC) in Africa, held in Accra in October 2023.

Ghanaian researchers' knowledge, control, and improvement of algorithms to make LCS work properly in African settings have led them to consider developing their own technology so as to be autonomous and no longer have to pay annual licence fees. This has led the Ghanaian physicists at UG to engage in the development and assembly of their own locally made LCS in order to reduce costs and have control of the technology. As we have clearly shown previously, all companies producing LCS around the world, whether American or Indian, depend on China to procure the main component. Therefore, Ghanaian scientists also plan to import the Plantower sensor from China and assemble it locally with a printed control board and a few cables (as shown in Figures 4.5a–c). Ghanaian scientists could, therefore, move beyond the pattern of invisible labour in science and innovation to redesign products and processes, which would mark self-determination and independence of their values and perspectives from Global North academics and start-up companies (Thambinathan and Kinsella 2021).

CONCLUSION: HOW ARE GHANAIAN SCIENTISTS SHIFTING AFRICA'S POSITION IN GLOBAL ATMOSPHERIC SCIENCE?

Ghana is one example in this collective book about how techno-scientific globalisation in the Global South is characterised in the form of serious constraints linked to international inequities. Indeed, Ghanaian atmospheric scientists and experts have always depended on international collaboration, external funds, and Global North instruments and tools to produce air pollution data and knowledge. Despite these hurdles and challenges, we have shown how the Ghanaian Physics Department's scientific team have managed to shape international collaborations since the mid-2000s, which have resulted in calibrating a fairly large number of LCS on which they depend to conduct atmospheric science research. This important validation work has resulted in a sufficiently robust and large body

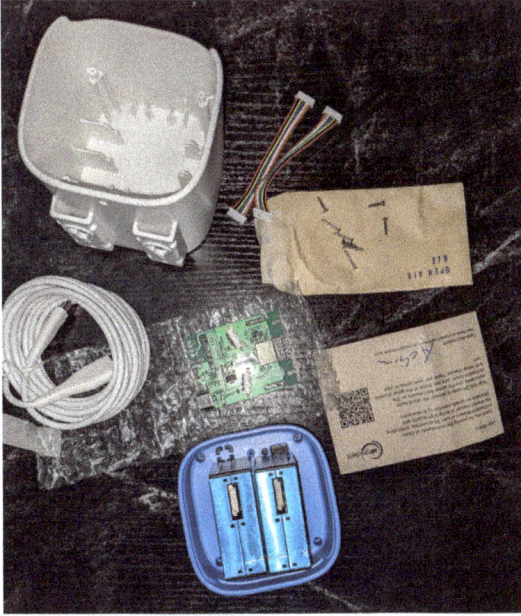

FIG. 4.5A Openaq low-cost sensor kit (Collins Gameli Hodoli)

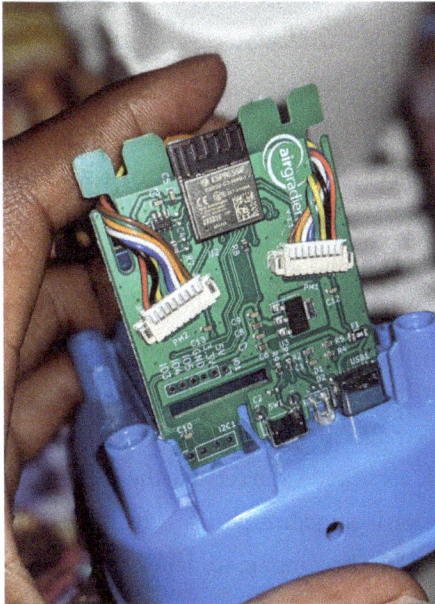

FIG. 4.5B Openaq printed control board (Collins Gameli Hodoli)

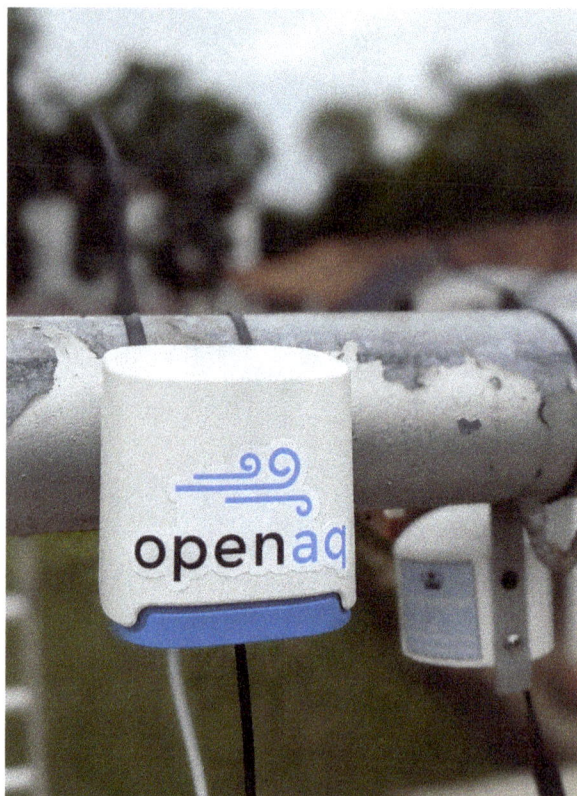

FIG. 4.5C Openaq sensor (Collins Gameli Hodoli)

of data, offsetting the fact that LCS still need to be scientifically evaluated and certified by regulatory agencies.

The Department of Physics at UG is thus becoming an African hub for the calibration and validation of LCS, which allows Ghanaian academics to produce scientific knowledge on air pollution. By appropriating tools and instruments from the Global North, they are creating alternative ways of doing air science to cope with constraints triggered by international inequities. Calibration technologies from the Global North and China enable them to use insights from the data they produce to develop legitimate expertise, which is, however, still determined from the perspective of the hegemonic academic actors. Nevertheless, Ghanaian scientists valorise their data by publishing their research results in internationally renowned scientific journals and making them visible to the global scientific community. Thus, Ghanaian scientists are coping with knowledge production inequities and reconfiguring the research ecosystem by shifting Africa's position in global atmospheric science.

They are also participating in innovation in a context of scarcity by adapting incoming technologies to local settings and appropriating these technologies with the ultimate goal of being able to assemble locally and control their own technology, which would be a marker of independence from the epistemological and commercial frameworks and ideologies coming from Europe and the United States.

These efforts, however, have taken almost two decades, and local scientists still have a long way to go before they are fully emancipated from the global academic hierarchy and international inequities. In the absence of a national research and innovation policy, Ghanaian researchers are still largely dependent on funding from academic bodies in the Global North. The case study we have presented in this chapter is certainly an encouraging jolt that is shaking up the established order in global socio-academic and techno-industrial structures. However, it also shows that the North American and Chinese companies that produce LCS and their components benefit from these power hierarchies by taking advantage of the free and invisible calibration labour carried out 'from below' by the little helpers – meaning local scientists – of global techno-scientific capitalism.

ENDNOTES

1 I refer here to the talk of Priyanka deSouza during the 1st episode of a podcast series titled *Decolonising Science* (https://www.colab.mit.edu/colabradio-more/decolonize-science-ep1 [accessed on 8 May 2023]).

2 Mordor Intelligence (2023) Analyse de la taille et de la part du marché des systèmes de surveillance de la qualité de lair – Tendances de croissance et prévisions (2024-2029) Source: https://www.mordorintelligence.com/fr/industry-reports/air-quality-monitoring-market (accessed 19th March 2025)

3 This contribution is the result of ongoing postdoctoral research I initiated in 2021 with the financial support of the Institut Francilien Recherche Innovation en Société (IFRIS) and am pursuing under the ANR-funded project Globalsmog (2022–2025). I am grateful to everyone who agreed to meet me and to be interviewed. I want to

thank especially my colleague Dr Allison Felix Hughes for our great collaboration; Emmanuel Appoh, former director of the Environmental Quality Department of the EPA-Ghana, for his trust to let me into EPA; and all the EPA team for their kindness and patience, and for agreeing to let me accompany them on surveillance tours. I also want to thank my academic colleagues (assistant researchers and PhD students) from the Department of Physics at UG for their welcome and friendship. A special thanks to my dear friend Dr Collins Gameli Hodoli for his inspiration and support. My gratitude also goes to my colleagues and friends of the seminar 'Technologization from Below' for their engagement and suggestions.

4 Dissertation defended in 2014 at the University of Ghana.

5 http://www.aqmd.gov/aq-spec [accessed on 8 May 2023].

6 https://www2.purpleair.com/products/purpleair-pa-ii?variant=40067691708513 [accessed on 8 May 2023].

7 Interview carried out on 7 June 2022 on Zoom with the VP, Business Development and Partnerships, at Clarity Movement Co.

8 https://aqicn.org/sensor/pms5003-7003/fr/ [accessed on 8 May 2023].

9 Rueda et al., *Environ Sci Technol Lett* 10 (2023): 247.

REFERENCES

Alli, A. S., S. N. Clark, A. Hughes, and others, 'Spatial-Temporal Patterns of Ambient Fine Particulate Matter (PM2.5) and Black Carbon (BC) Pollution in Accra', *Environmental Research Letters*, 16 (2021): 074013 https://doi.org/10.1088/1748-9326/ac074a.

Allen, B., 'Strongly Participatory Science and Knowledge Justice in an Environmentally Contested Region', *Science, Technology, & Human Values*, 43 (2018): 947–71 https://doi.org/10.1177/0162243918758380.

Arku, R. E., J. Vallarino, K. L. Dionisio, and others, 'Characterizing Air Pollution in Two Low-Income Neighbourhoods in Accra, Ghana', *Science of the Total Environment*, 402 (2008): 217–31 https://doi.org/10.1016/j.scitotenv.2008.04.042.

Bangham, J., X. Chacko, and J. Kaplan, *Invisible Labour in Modern Science* (Rowman & Littlefield, 2022).

Bhaduri, S., 'Frugal Innovation by "The Small and the Marginal": An Alternative Discourse on Innovation and Development', [Conference session], *Prince Claus Chair Inaugural Lecture*, ISS, The Hague, the Netherlands, 23 May 2016 https://doi.org/10.13140/RG.2.1.1290.9682.

Clark, S. N., A. S. Alli, M. Brauer, and others, 'High-Resolution Spatiotemporal Measurement of Air and Environmental Noise Pollution in Sub-Saharan African Cities: Pathways to Equitable Health Cities Study Protocol for

Accra, Ghana', *BMJ Open*, 10 (2020): e035798 https://doi.org/10.1136/bmjopen-2019-035798.

Davies, S. R., *Hackerspaces: Making the Maker Movement* (Polity Press, 2017).

deSouza, P., 'Air Pollution in Kenya: A Review', *Air Quality, Atmosphere & Health*, 13 (2020): 1487–95 https://doi.org/10.1007/s11869-020-00902-x.

Dionisio, K. L., R. E. Arku, A. F. Hughes, and others, 'Air Pollution in Accra Neighbourhoods: Spatial, Socioeconomic, and Temporal Patterns', *Environmental Science & Technology*, 44 (2010a): 2270–76 https://doi.org/10.1021/es903276s.

Dionisio, K. L., M. S. Rooney, R. E. Arku, and others, 'Within-Neighbourhood Patterns and Sources of Particle Pollution: Mobile Monitoring and Geographic Information System Analysis in Four Communities in Accra, Ghana', *Environmental Health Perspectives (Online)*, 118 (2010b): 607–13 https://doi.org/10.1289/ehp.0901365.

Gabrys, J., H. Pritchard, and B. Barratt, 'Just Good Enough Data: Figuring Data Citizenships through Air Pollution Sensing and Data Stories', *Big Data & Society*, 3 (2016) https://doi.org/10.1177/2053951716679677.

Garrocq, J. B., and S. Parasie, 'Peut-On Redistribuer la Surveillance de la Qualité de l'Air? Une Enquête sur les Métrologies à l'Oeuvre dans un Concours Institutionnel d'Évaluation de Micro-Capteurs', *Revue d'Anthropologie des Connaissances*, 16 (2022) https://doi.org/10.4000/rac.29150.

Giordano, M. R., C. Malings, S. N. Pandis, and others, 'From Low-Cost Sensors to High-Quality Data: A Summary of Challenges and Best Practices for Effectively Calibrating Low-Cost Particulate Matter Mass Sensors', *Journal of Aerosol Science*, 158 (2021): 105833 https://doi.org/10.1016/j.jaerosci.2021.105833.

Hagan, D. H., and J. H. Kroll, 'Assessing the Accuracy of Low-Cost Optical Particle Sensors Using a Physics-Based Approach', *Atmospheric Measurement Techniques*, 13 (2020): 6343–55 https://doi.org/10.5194/amt-13-6343-2020.

Harzing, A., *The Publish or Perish Book: Your Guide to Effective and Responsible Citation Analysis* (Tarma Software Research PTY Ltd., 2010).

Jack, D. W., K. P. Asante, B. J. Wylie, and others, 'Ghana Randomised Air Pollution and Health Study (GRAPHS): Study Protocol for a Randomised Controlled Trial', *Trials*, 16 (2015): 420 https://doi.org/10.1186/s13063-015-0930-8.

Katoto, P., L. Byamungu, A. S. Brand, and others, 'Ambient Air Pollution and Health in Sub-Saharan Africa: Current Evidence, Perspectives and a Call to Action', *Environmental Research*, 173 (2019): 174–88 https://doi.org/10.1016/j.envres.2019.03.029.

Landrigan, P. J., R. Fuller, N. J. R. Acosta, and others, 'The Lancet Commission on Pollution and Health', *The Lancet*, 391 (2017): 462–512 https://doi.org/10.1016/S0140-6736(17)32345-0.

Lewis, A. C., J. D. Lee, P. M. Edwards, and others, 'Evaluating the Performance of Low-Cost Chemical Sensors for Air Pollution Research', *Faraday Discussions*, 189 (2016): 85–103 https://doi.org/10.1039/C5FD00201J.

McFarlane, C., R. Garima, C. Malings, and others, 'Application of Gaussian Mixture Regression for the Correction of Low-Cost PM2.5 Monitoring Data in Accra, Ghana', *ACS Earth Space Chem*, 5 (2021): 2268–79 https://doi.org/10.1021/acsearthspacechem.1c00217.

Moizard-Lanvin, J., 'Sélectionner et Agréger les Ignorances', *Revue d'Anthropologie des Connaissances*, 15 (2021) https://doi.org/10.4000/rac.25114.

Morawska, L., P. K. Thai, X. Liu, and others, 'Applications of Low-Cost Sensing Technologies for Air Quality Monitoring and Exposure Assessment: How Far Have They Gone?', *Environment International*, 116 (2018): 286–99 https://doi.org/10.1016/j.envint.2018.04.018.

Negi, R., and A. Ranjan, 'Delhi's Fight Against Air Pollution Could Get a Boost if Science Was Decolonised', *Scroll.in*, 11 November 2020 https://scroll.in/article/976361/delhis-fight-against-air-pollution-could-get-a-boost-if-science-was-decolonised.

Ottinger, G., 'Buckets of Resistance: Standards and the Effectiveness of Citizen Science', *Science, Technology, & Human Values*, 35 (2010): 244–70 https://doi.org/10.1177/016224390933712.

Parasie, S., and F. Dedieu, 'À Quoi Tient la Crédibilité des Données Citoyennes? L'Institutionnalisation des Capteurs Citoyens de Pollution de l'Air en Californie', *Revue d'Anthropologie des Connaissances*, 13 (2019): 1035–62 https://doi.org/10.3917/rac.045.1035.

Pourraz, J., 'Production and Circulation of Local Knowledge About Air Pollution and Health Effects in Ghana', in K. W. Fomba, B. Tchanche Fankam, A. Mellouki, D. M. Westervelt, and M. R. Giordano, eds., *Advances in Air Quality Research in Africa* (Cham: Springer, 2024), pp. 421–39 https://doi.org/10.1007/978-3-031-53525-3_24.

Pritchard, H., and J. Gabrys, 'From Citizen Sensing to Collective Monitoring: Working Through the Perceptive and Affective Problematics of Environmental Pollution', *GeoHumanities*, 2 (2016): 354–71 https://doi.org/10.1080/2373566X.2016.1234355.

Pritchard, H., J. Gabrys, and L. Houston, 'Re-Calibrating DIY: Testing Digital Participation Across Dust Sensors, Fry Pans and Environmental Pollution', *New Media & Society*, 20 (2018): 1–20 https://doi.org/10.1177/1461444818777473.

Rooney, M. S., R. E. Arku, K. L. Dionisio, and others, 'Spatial and Temporal Patterns of Particulate Matter Sources and Pollution in Four Communities in Accra, Ghana', *Science of the Total Environment*, 435–36 (2012): 107–14 https://doi.org/10.1016/j.scitotenv.2012.06.077.

Rueda, E., E. Carter, C. L'Orange, C. Quinn, and J. Volckens, 'Size-Resolved Field Performance of Low-Cost Sensors for Particulate Matter Air Pollution', *Environmental Science & Technology Letters*, 10(3) (2023): 247–53 https://doi.org/10.1021/acs.estlett.3c00030.

Sahu, R., K. K. Dixit, S. Mishra, and others, 'Validation of Low-Cost Sensors in Measuring Real-Time PM10 Concentrations at Two Sites in Delhi National Capital Region', *Sensors*, 20 (2020): 1347 https://doi.org/10.3390/s20051347.

Thambinathan, V., and E. A. Kinsella, 'Decolonising Methodologies in Qualitative Research: Creating Spaces for Transformative Praxis', *International Journal of Qualitative Methods*, 20 (2021): 1–9 https://doi.org/10.1177/16094069211014766.

Wooldridge, A., 'The World Turned Upside Down: A Special Report on Innovation in Emerging Markets', *The Economist*, 15 April 2010 http://www.economist.com/node/15879369.

World Health Organization, *WHO Global Air Quality Guidelines: Particulate Matter (PM2.5 and PM10), Ozone, Nitrogen Dioxide, Sulfur Dioxide, and Carbon Monoxide* (World Health Organization, 2021).

Zhou, Z., K. L. Dionisio, T. G. Verissimo, and others, 'Chemical Composition and Sources of Particle Pollution in Affluent and Poor Neighbourhoods of Accra, Ghana', *Environmental Research Letters*, 8 (2013): 044025 https://doi.org/10.1088/1748-9326/8/4/044025.

5

AFFIRMING PHARMACEUTICAL SOVEREIGNTY: TECHNOLOGY TRANSFER AGREEMENTS AND VACCINE GEOPOLITICS DURING A GLOBAL HEALTH EMERGENCY

Koichi Kameda, Denise Pimenta, and Gustavo Matta

FIG. 5.1 'Vacinas para Todos Já!!!' / 'Vaccines for all now!!!'. Teresópolis, State of Rio de Janeiro, January 2021 (Koichi Kameda)

IN JANUARY 2021, BRAZIL WAS ONE OF THE COUNTRIES WORST AFFECTED by COVID-19. Suffering 200,000 deaths, it was second only in casualties to the United States (Instituto Butantan 2021). On 17 January, the Brazilian health regulatory agency, ANVISA, approved the emergency use of AstraZeneca/ University of Oxford and Sinovac's vaccines (ANVISA 2021). Both were made available as part of the partnerships with two Brazilian public laboratories: Bio-Manguinhos/Fiocruz (hereafter Bio-M) and Butantan Institute. These partnerships, which involved the circulation of knowledge and vaccine shots, national regulatory work, and public and philanthropic funding, were implemented to enable a vaccine campaign in global and national geopolitical scenarios normally unfavourable to public health responses in the developing world. By the following January 81% of the targeted population had been vaccinated.[1]

One might recall the erratic management of the Brazilian response to COVID-19 when the virus first arrived in the country, particularly the denialism that characterised President Jair Bolsonaro's engagement with the pandemic. But this mismanagement was present in local as well as federal government, with its effects felt mainly by the regions and populations in more vulnerable situations (Ortega and Orsini 2020). Despite being one of the favourite targets of the president and his followers' attacks, vaccines were in great demand throughout the whole population, being seen as tools that could reduce deaths and normalise the situation in the country (Peixoto et al. 2023). And indeed, vaccines became key actors in fighting the COVID-19 pandemic.

The development of COVID-19 vaccine innovations has been remarkable, benefiting from scientific advances in response to earlier global health emergencies (Lurie et al. 2020), as well as a combination of funding boosts, regulatory flexibility, and the introduction of 'next-generation' platforms such as RNA and viral vector-based vaccines, in co-existence with classic technologies (van Riel and de Wit 2020).[2] However, this technological development has gone hand in hand with stark inequalities in access to COVID-19 vaccines between countries. The cooperation between nations expected in such a situation gave way to technological nationalism, with the wealthiest and most industrialised countries racing to secure most vaccine doses, many times greater than those needed to immunise their populations (Jensen et al. 2022). Meanwhile, the monumental failure of the COVID-19 Vaccine Global Access (COVAX) Facility, the global health initiative supposed to make shots equally available in the world, also played a role in this inequality. The pandemic called into question the mainstream pharmaceutical belief in 'globalisation from the North,' which was implicit in the idea that in such emergencies 'magic bullets' like vaccines would be developed in the North to use in the South (Medina et al. 2014; Pollock 2019). The hoarding of vaccines by the Global North also highlights that during the pandemic, Northern countries halted the globalisation process through 'technological nationalism', effectively undermining international cooperation through the World Health Organisation's COVAX scheme. In contrast, Southern countries continued to push for more globalisation to gain access to knowledge and technologies.

The uneven geographies of pharmaceutical production and access can be seen as a result of the globalisation of the norms related to intellectual property rules and regulatory harmonisation intended to protect technologies and actors from the North, kicking away the ladder from actors in the South (Chang 2002) and consolidating the central role of Big Pharma companies (Sunder Rajan 2017). In this context of pressing inequalities, actors from below might look for creative ways to circumvent the effects of globalisation from above and establish new relationships with actors in the South. Indeed, industrial actors in the Global South already contribute most of the vaccines made available in much of the world today: China and India host 31% of global vaccine manufacturers (WHO 2023). China, India, and Russia have played a crucial role in supplying COVID-19 vaccine doses to low- and middle-income countries (LMICs).[3] China alone has supplied nearly 1.3 billion doses to LMICs – more than the global COVAX Facility (Wang 2022). China is also a significant supplier of drug substances (active pharmaceutical ingredients, APIs) to other countries' pharmaceutical companies for vaccines and antibiotics (Zhang and Bjerke 2023). APIs are a crucial part of vaccines, representing the components made from viruses or bacteria (sometimes also called 'antigens'). They are responsible for the vaccine's role in challenging the immune system in such a way as to induce it to fight the disease.

What is the space for Southern countries to develop their technological enterprises within an unequal global scenario marked by old and new dependency relations among countries? How do these national enterprises become feasible, interact with forms of globalisation from the North and the emerging South, and contribute to alternative forms of globalisation? To address these questions, this chapter analyses the manufacturing of COVID-19 vaccines in Brazil. One key concept is pharmaceutical sovereignty, which involves the state's ability to implement national health policies with regards to promoting access to essential health technologies, from medicines to vaccines and diagnostics. Pharmaceutical sovereignty also comprises the existence of domestic technological and industrial capacity to address local health needs.[4] The affirmation of sovereignty forms has been mainly analysed in the context of developing countries' resistance to the compulsory sharing of biological materials in the context of the fight against previous public health emergencies (Fearnley 2020). However, this has happened without any guarantee of access to the benefits of technological development. In some cases – as with the zika epidemics – this reluctance to share knowledge and technology has had a marked negative impact on timely development and innovation (Kameda et al. 2021; Kelly et al. 2020). We draw on how this bio-sovereignty (van Wichelen 2023) has become increasingly understood in terms of bio-production sovereignty, due to its connection with the right of countries to produce their own health biotechnologies (Kameda 2021). We argue that the increasing production and development role of Southern countries transcends the standard attribution of the South merely as a site of extraction and demonstrates that Global South countries are able to develop technologies that best fit their own needs and independently address national health policies (Pollock 2019). Despite the restrictive norms imposed by global pharmaceutical capitalism, developing countries are still able to exercise sovereignty, be it through alliances that enable the use of patent flexibilities to produce antiretroviral generic drugs (Biehl 2007; Hayden 2007) or material transfer agreements (van Wichelen 2023).

This chapter shows how pharmaceutical sovereignty was at stake in Brazil during the COVID-19 pandemic through the analysis of the production of COVID-19 vaccines. The vaccine sector, though historically concentrating on states' political efforts to achieve production self-reliance, has changed in recent decades. These changes, prompted by a global trend towards state disengagement from vaccine production, has increasingly resulted in their com-modification (Blume 2017; Blume and Baylac-Paouly 2021). In the opposite direction, Brazil has established a public vaccine industry due to a longstanding drive to reduce its dependence on imports of health goods, a commitment reaf-firmed throughout the country's history following episodes that exposed its technological vulnerability (Benchimol 2017). It relies on a successful model of technology transfer partnerships to obtain advanced technologies that interest the public health system.

During the COVID-19 pandemic, the two major Brazilian public vaccine manufacturers engaged in the production of COVID-19 vaccines through part-nerships: Bio-M – linked to the Oswaldo Cruz Foundation (Bio-M/Fiocruz) – partnered with the British multinational AstraZeneca, which had obtained an exclusive license from the University of Oxford to explore its technology. Meanwhile, the Butantan Institute partnered with the Chinese company Sinovac. Despite the national context of virus denialism and politicisation of vaccines, the Brazilian producers were able to implement this model (Medeiros et al. 2022) thanks to their pre-existing production capacity, their ability to source fund-ing from the national public and private sectors, and regulatory support from the National Health Regulatory Agency, or ANVISA (Fonseca et al. 2023). By analysing the partnership between Bio-M and AstraZeneca, this chapter con-tributes to the understanding of the links between the exercise of bioproduction sovereignty and the literature on globalisation from below, showing nuances in the power relations in technological partnerships between North and South, but also how national technological projects need to adapt to a new scenario of influential Southern players.

The chapter is organised in four sections. After a first section that outlines the particularity of the Brazilian vaccine manufacturing model, the chapter will analyse the partnership of Fiocruz and AstraZeneca to locally manufacture the University of Oxford's COVID-19 vaccine to discuss how the Brazilian actors exercised pharmaceutical sovereignty during a global health emergency. The second and third sections discuss the negotiations to conclude the agreements and purchase vaccine ingredients and knowledge 'transfer'. They indicate the agency of Brazilian actors in pushing for the technological partnership. By purchasing AZ vaccine components and manufacturing knowledge, Brazilian actors indeed supported the technological development of the Northern vaccine. However, the negotiation of contractual terms such as the freedom to use transferred platforms remains a challenge. In the fourth section, the chapter analyses the exercise of sovereignty vis-à-vis the growing dependency of the global pharmaceutical value chain on Chinese and Indian companies. Behind the AZ vaccine are companies from the South that supported the vaccine development initiative in Brazil in a context of scarcity, revealing new asymmetric relations of power. In this scenario, Brazil exerted its sovereignty through diplomatic negotiations over the vital pharmaceutical inputs, but also through work related to the logistics of vaccine importation (establishing plans and partnerships with private and public stakeholders, organising international transport, negotiating customs permissions, the technical work of quality and temperature control, etc.) required to assure vaccine components arrived promptly in Brazil during a global health crisis. The chapter concludes by reflecting on how the multidirectionality of technoscientific globalisation interrelates and affects pharmaceutical sovereignty and new forms of South–South circulation of vaccine technologies.

We mobilised the following methods for data collection: interviews with Fiocruz staff conducted between August and November 2022;[5] consulting open-access presentations, contracts, and documents; and consulting documents made public during the Parliamentary Commission of Inquiry into the COVID-19 Pandemic,[6] as well as an analysis of how the Brazilian press covered the vaccine rollout.

PHARMACEUTICAL SOVEREIGNTY? TECHNOLOGY PARTNERSHIPS AND PUBLIC VACCINE MANUFACTURING

Vaccine technology institutions in Brazil

One of the specificities of the Brazilian pharmaceutical industry is that it includes a set of 'official' pharmaceutical laboratories. These are laboratories under the remit of federal or state governments, universities, and the army. In 2014, 18 official laboratories were active in four Brazilian regions (Hasenclever et al. 2018; Oliveira et al. 2014). In the vaccine sector, two official laboratories, Fiocruz and Butantan, are the leading local vaccine producers and suppliers of the National Immunisation Programme (in Portuguese, Programa Nacional de Imunizações, PNI) of the National Healthcare System (in Portuguese, Sistema Único de Saúde – SUS) – PNI accounts for 90% of the demand for human vaccines in the country (Gadelha et al. 2020).[7] Even though primarily focusing on the SUS' vaccine demands, Bio-M/Fiocruz and Butantan are also relevant actors in global health, being amongst the biggest vaccine players worldwide (WHO 2023), with WHO prequalified yellow fever (Bio-M/Fiocruz) and trivalent influenza (Butantan) vaccines.[8]

Fiocruz is part of the Ministry of Health and comprises 16 technical and scientific units 'focused on teaching, research, innovation, assistance, technological development and extension in the health field'.[9] It has a presence in 10 Brazilian states and an office in Maputo, Mozambique. The core purpose of Fiocruz's vaccine-manufacturing activities is to supply the PNI, which was launched in 1973 to control major infectious disease epidemics in the country. The PNI established several compulsory national vaccination schemes against infectious diseases. The PNI model was inspired by the success of the Smallpox Eradication Campaign, which was sustained by mass vaccination initiatives and involved robust monitoring and control. In this programme a range of vaccines were being purchased by the state and freely distributed to the population. The Brazilian PNI was created during one of the military dictatorship governments and in a context marked by the Brazilian population's vulnerability to preventable diseases, particularly when a meningococcal meningitis epidemic, as well as polio and measles outbreaks, hit the country (Buss et al. 2005).

While Fiocruz's history in the production of biologicals can be traced back to the early twentieth century,[10] since its establishment in its current form in the 1970s, the Foundation's pharmaceutical and biological manufacturing activities have been concentrated in its two pharmaceutical laboratories, Far-Manguinhos and Bio-M, respectively. Bio-M, set up in 1976, is the unit in charge of manufacturing vaccines, diagnostic components, and biopharmaceuticals. When Bio-M was created, its industrial facilities were at first precarious.[11] The laboratory focused its initial efforts on providing infrastructure for the required industrial activities and the acquisition of vaccine technology. It decided to invest in the strategy of negotiating technology transfer agreements – a strategy which Bio-M had already employed for its yellow fever vaccine (Löwy 2006). This was seen as a way to 'rapidly introduce vaccines on the market while learning and strengthening [the laboratory]' (interview with Bio-M's representative, 2022). This strategy enabled the Brazilian organisation to begin vaccine production and diversify its portfolio.

Vaccine production models and technology transfer agreements

To get started, in 1976 and 1980 Bio-M negotiated agreements with not-for-profit institutions in France and Japan to obtain the technologies to produce vaccines for meningitis, poliomyelitis, and measles.[12] These technology transfer agreements were signed as part of cooperation frameworks. Admittedly, commercial and trade interests were at stake in these agreements, albeit indirectly. Nevertheless, the terms were highly favourable to Bio-M, considerably advancing its goal of mastering the manufacturing of the measles and polio vaccines. This took place in a context where the production of vaccines was not of interest to multinational pharmaceutical companies.

Given the dramatic consequences of PNI product shortages, the Brazilian state decided to invest in the national public production of vaccines. One such shortage in particular prompted the government to invest in expanding vaccine production in the public sector. In the 1980s, a branch of the multinational Syntex, which was the government's leading supplier of the DTP (diphtheria, tetanus, and pertussis) vaccine, as well as the largest snake antivenom manufacturer in Brazil, saw the commercialisation of its products suspended over quality issues. Rather than invest in enhancing the quality control of its vaccines, the company decided to shut down its production lines in Brazil, which led to a shortage of the products used by the health services (Ponte 2007). Moreover, there were problems with the supply of other vaccines (tetanus toxoid, rabies vaccine, and BCG). To remedy the situation, in 1985, during Brazil's democratic opening, the government created the National Immunobiological Self-Sufficiency Programme (Programa de Auto-Suficiência Nacional em Imunobiológicos,

PASNI). The programme focused most of its investments on the modernisation and building of facilities for the laboratories that were part of the programme, including Bio-M and Butantan.

Subsequently, starting in the late 1990s, Bio-M faced a major existential crisis due to the national political and economic situation. To solve this crisis, which almost caused the institution to shut down, Bio-M implemented a set of administrative and organisational reforms. In particular, given the urgent need to update its scientific and technological platforms and infrastructures, the laboratory adopted a policy of investing in 'high value-added' products that it would seek to manufacture by negotiating technology transfer agreements. Bio-M targeted the production of more expensive technologies that the Brazilian state intended to purchase. This resulted in the establishment of technology transfer partnerships, mostly with multinationals.

Productive development partnerships (PDPs)

The first partnership of this kind consisted of an agreement with the Belgian company SmithKline (now GlaxoSmithKline) in 1998 to transfer the Hib (Haemophilus influenzae type B) vaccine. Through this agreement, SmithKline was to receive access to the Brazilian public market for five years and receive 65% of the profits from selling the product to PNI during this time, with the remaining 35% going to Bio-M. Following the end of these five years formally recognised as the duration of the technology transfer, Bio-M was to acquire the production capacity for the vaccine, receive the total revenue from its sale, and gain the right to commercialise the nationalised product in the Mercosur trade bloc, the economic community of South American countries. The state would also benefit by purchasing the vaccine at the price set by the Revolving Fund mechanism of the Pan American Health Organisation (PAHO).[13] According to Ponte (2007), partnering with Fiocruz was also valuable for SmithKline, as the public laboratory had invested in expanding its capacity for fill-finish steps. Fill-finish is one of the phases of the productive process of making a vaccine available.[14] In the fill-finish phase, the bulk vaccine is transferred into glass bottles, sealed, lyophilised if applicable, labelled, and packed. Bio-M inaugurated its fill-finish centre in 1998, which played a crucial role in securing the agreement with SmithKline since lyophilisation constituted an expensive component in the manufacturing of vaccines, and the Belgian company had a bottleneck in that production step. In order to supply the vaccine in the quantities required by the government, the multinational would need to invest directly in various production steps in the country. Thus, the partnership offered the multinational easier access to the Brazilian market. At the same time, for Fiocruz, the localisation of Hib represented an unprecedented revenue increase that enabled it to invest in the unit's physical and administrative infrastructure needs and research to improve and develop products (Ponte 2007).[15]

In the following years, Fiocruz established new agreements leveraging public purchasing power to attract partners to engage in technology transfer agreements. These pertained to vaccines, rapid diagnostic tests, and biopharmaceuticals. Additionally, in 2001, Fiocruz obtained the WHO prequalification of its yellow fever vaccine, becoming one of just four suppliers of this vaccine worldwide.[16] With the incorporation of new products into its portfolio and the growth of yellow fever vaccine exports following the WHO certification, the Fiocruz vaccine unit, Bio-M, saw its revenue increase by 6,300% in 2004 – from BR$4 million (Brazilian reals) to BR$280 million (Ponte 2007).

As seen in this section, Bio-M has built a pharmaceutical production model characterised by public production based on partnerships with private proprietary companies to transfer technology and production know-how. In contrast to the experience of Fiocruz's other production laboratory, Far-Manguinhos, which challenged patents for HIV medicines in order to enable it to produce generics,[17] Bio-M is engaged in a partnership model (Cassier and Correa 2018). This model 'evolved' from international cooperation with for-profit institutions to the public-private partnership model with for-profit institutions, including 'Big Pharma', but also with some partners in the South (for example, Cuba), to rapidly introduce new technologies and platforms to its portfolio. From this model, Bio-M continues to primarily serve the Brazilian public sector (and, in a subsidiary way, other LMIC countries and international organisations). However, it has been increasingly establishing partnerships for 'high value-added' products, that is, those that cost more for the Brazilian government.

Even if this public model has no profit orientation, the product is not sold at a cost price, and the 'excess' is reinvested in research and infrastructure expansion. Moreover, this model of TT partnerships has been generalised as a significant device of industrial policies for the pharmaceutical sector since the 2000s by the successive Workers' Party governments (Chaves et al. 2016).[18] These partnerships, named productive development partnerships (PDPs), target the local production of health technologies that are regularly named as a priority by the Ministry of Health in the sense of making them freely available to the population through the public health system. At the same time, they are intended to reduce the deficit in the trade balance by reducing imports, which is behind the current approach of bolstering the country's 'healthcare economic-industrial complex'.

Even though the Bio-M model (and the current health industrial complex framework) relies on technology and knowledge transfer from other companies, this is not a linear relationship that reproduces the mainstream idea of globalisation from the North. The state and public laboratories are behind the choice of technologies, and they trigger the negotiation of partnerships with vaccine owners. Access to the vast Brazilian public market of health technologies is conditioned by access to the knowledge that would enable faster creation of capacity to manufacture and make the technologies available at SUS, a process usually referred to by actors on the ground as 'nationalisation'. In turn, the companies sell their product exclusively to the government during the technology transfer, which can last five to ten years (Cassier and Correa 2018).

This strategy of partnerships has enabled the Brazilian public sector to 'capture' new vaccine technologies to add to its portfolio, reducing its dependence on imports.[19] Ironically, the country's trade balance has gradually deteriorated with the importation of new vaccines as part of the PNI. Notably, it reached an average annual trade deficit of US$664 million between 2015 and 2018 (Gadelha et al. 2020). This trade deficit reveals the limits of this technology transfer–based approach in reducing the country's technology dependence. The choice of a model of pursuing technology independence through partnerships, instead of more subversive approaches of reverse engineering for copying a medicine, as happened with HIV pharmaceuticals, is usually justified by the fact that patents are not considered the main barrier to the production of biologicals like vaccines; instead, know-how is thought of as the most essential requirement. Moreover, challenging patents in the post–TRIPS Agreement scenario comes with a political cost that is too high for governments in the South. Partnerships with proprietary companies seem like a more viable strategy. However, this approach of pursuing autonomy through partnerships that involve the importation of a whole sociotechnical system reinforces a vicious circle, where more 'capture' leads to more technological dependency, feeding the globalisation process from the North.

TECHNOLOGICAL LANDSCAPE AND THE 'SELECTION' OF THE OXFORD-ASTRAZENECA VACCINE

Before Bio-M became involved in vaccine manufacturing to fight COVID-19, the laboratory and the Ministry of Health had been monitoring the global landscape of technological development surrounding SARS-CoV-2 vaccines since March 2020. Bio-M had been discussing the vaccine technologies with the proprietary companies as a step towards establishing a partnership. At that point of the pandemic, no vaccines developed by Brazilian institutions were at an advanced stage, justifying pursuing the previous strategy of negotiating a partnership with foreign companies.

Critically, the company had to agree to transfer technology to set up a partnership following the Brazilian institution's strategy. Pfizer-BioNTech, whose RNA vaccine platform was introduced during the pandemic and had the potential of being applied to numerous immunisation and treatment purposes, conducted clinical trials in Brazil and was interested in selling the shots to the country but refused to engage in a partnership involving knowledge transfer. In Bio-M's discussions with other companies,[20] only two agreed to transfer technology: Sinovac and AstraZeneca. Both companies implemented manufacturing models that involved contracting with manufacturers in the South to expand production capacity. Just as most developers based in high-income countries (HICs) have partnered with manufacturers based in other HICs, companies based in middle-income countries (MICs – China, Cuba, India, Kazakhstan,

Russia, and Vietnam) also primarily selected partners in MICs to expand their manufacturing capacity (Global Health Centre 2021).

Sinovac developed one of the four Chinese vaccines with the most significant international footprint – the other companies were Sinopharm-Beijing, Sinopharm-Wuhan, and CanSino. Sinovac's vaccine, CoronaVac, was based on the classic inactivated virus platform. Ultimately, the Chinese company chose to partner with the Butantan Institute, an initiative considered co-development by Butantan given its involvement in conducting phase 3 of the vaccine clinical trials in the country.[21] However, the absence of infrastructure necessary to produce inactivated virus platforms – Level 3 biosafety laboratories, non-existent in the country at that time (interview with Bio-M director, 2022) – prevented the total production of the vaccine in Brazil, a situation that necessitated continuous importation of pharmaceutical ingredients from Sinovac. Indeed, as discussed later, Chinese companies played a role in assuring the Fiocruz and Butantan technological initiatives for COVID-19, indicating an emerging technological globalisation that can sustain national policies in the South, together with market expansion and the creation of new forms of dependency.

In its turn, the vaccine manufactured by the UK-based pharmaceutical AstraZeneca (AZ) was a new vaccine platform based on chimpanzee adenovirus (ChAdOx1 nCoV-19) that was developed by University of Oxford researchers and a company they created, Vaccitech. The Oxford vaccine development took place in an international network that initially had no plans to exclusively license the technology to any one company, an approach that changed under pressure from Bill Gates (Garrison 2020). However, the company pursued the commitment of not making a profit during the pandemic, charging only production and distribution costs. Because the large pharmaceutical company had never operated in the vaccine industry, it created a delocalised model of manufacturing, with agreements to produce API or fill and finish activities in 14 locations, but, in contrast to other HIC-based vaccine developers, AZ established manufacturing partnerships with 5 MICs (Brazil, China, India, Russia and Thailand) (Global Health Centre 2021).

The AstraZeneca technology transfer was effectively kicked off by a Brazilian researcher in charge of the vaccine's development at the company, Sue Costa

Clemens. She first approached the Brazilian Ministry of Health on 8 May 2020 as part of the preparation of Phase 3 clinical trials in Brazil, to determine the vaccine's safety, efficacy, and immunogenicity. After technical meetings involving AstraZeneca Brazil and the Embassy of the United Kingdom, the company expressed its interest in partnering with Bio-Manguinhos and agreed to transfer the vaccine technology, a decision that was formally made on 22 June 2020 (Clemens 2021).[22]

Aside from the fact that the Oxford-AZ vaccine was at the most advanced technological development stage at that time – in Phase 2-3 – and had just started Phase 3 clinical trials in Brazil, another benefit of the agreement was the vaccine platform itself, according to Bio-M. It represented a new addition to the laboratory's vaccine portfolio, with the possibility of developing future uses. Moreover, the similarity between the manufacturing processes of the Oxford vaccine and those of a biopharmaceutical previously introduced in Bio-M's manufacturing routine would facilitate the transfer. Bio-M had added two biological medicines – the recombinant human erythropoietin (alfapoetina) and the interferon alpha-2b – to its portfolio through partnerships with two Cuban institutions, the Centro de Imunologia Molecular and Heber Biotech, in 2004 and 2008. The acquired alfa-poetina production capacity, in particular, with the use of 'Chinese hamster ovary cell' (CHO) cultures using bioreactors, would facilitate the technology transfer.[23] Bio-M decided to use the industrial area previously dedicated to its other biopharmaceutical, interferon, for COVID-19 vaccine manufacturing, since the former's production had been interrupted after its replacement by more effective medicines in hepatitis C treatment (treatments based on 'direct-acting antiviral agents').

Even though most of Bio-M's partnerships are with Big Pharma companies and institutions from the North, this episode shows that Southern institutions also own strategic technologies that can make them equally essential partners of Brazilian technological enterprises. Furthermore, the partnership with Cuban institutions counterbalances the criticism of the technology transfer–based model of pharmaceutical capacity creation by indicating that this strategy can bring benefits beyond the development of one specific product, installing infrastructure and expertise that can later benefit other technological projects. The existing production capacity for erythropoietin informed the decision to go for the vaccine partnership with AstraZeneca and indicates Bio-M's agency in the partnership. It contributes to questioning the assumption in global health discourses of the South as a place dispossessed of production capacity. It reinforces the argument that the primary barrier to more institutional manufacturing in the South is the assetisation of IP by the North (Birch 2020) rather than the unavailability of previous infrastructure. But how did the negotiations proceed in practice?

THE NEGOTIATION OF LEGAL INSTRUMENTS TO ENABLE TECHNOLOGY TRANSFER AND VACCINE LOCALISATION

The pandemic forced Bio-M to engage in negotiations to secure a technology that was still under development. Although the AZ vaccine was the most advanced vaccine candidate when the partnership was established – it had begun Phase 3 clinical trials at the time of its selection for the Bio-M partnership – it still carried the risk of revealing data that would upend the venture. For Bio-M and the Brazilian government, this meant investing upfront in technology despite the risk of it failing. This type of instrument, namely the leveraging of public procurement to spur innovation in different areas, has been employed in Europe and the United States for many years (Edquist and Zabala-Iturriagagoitia 2012). However, the Brazilian government had never implemented such a tool for vaccine procurement – the strategy of Bio-M, as part of the federal government, had so far involved negotiating the purchase and transfer of technology already available on the market and registered with the regulatory agency. A specific technological contract was then adopted: the 'encomenda tecnológica' (ETEC, or technology purchase order), a type of contract first introduced by the Brazilian Innovation Law of 2004 and modified by the 2016 Scientific and Technological Framework (Marco da C&T). This ETEC instrument allows the state to acquire 'an R&D venture to find a solution not available on the market for a specific application'.

> Encomenda tecnológica is a process that allows you to pay in advance for a project that may not work out [...] everyone wanted to sell and not put

their own money in, these companies, big multinationals wanted to sell, the US put much money in, the European countries, in the governments that have more flexibility, let us say, put the money from the pharmaceutical companies and they sold vaccines that did not exist yet, that may not even exist. […] And it was still under clinical study, and many people know that many vaccines are rejected during the clinical study phase. (Interview with Bio-M director, 2022)

This legal instrument was employed for the first time for a vaccine enterprise in Brazil. It enabled technological globalisation from a Northern vaccine but in the terms defined by Bio-M and its purpose of nationalising the imported technology. The aforementioned interview also sheds light on how a Southern state can support technological development in the North. Moreover, this new instrument contributed to risk sharing of an innovation that was still in development, despite its advanced stage, and had the risk of not reaching the market.

The stated intention to agree did not make the negotiation of the partnership conditions any less critical. It required the involvement of lawyers from different Fiocruz departments (Bio-M, its technological management coordination department, and the office of the lawyer of Fiocruz) as well as officials of the Foundation to face critical aspects of the negotiation, as described by a lawyer who took part in the process.

A first critical aspect was that these negotiations had to consider the urgent need to secure drug substances in a context marked by competition for vaccine doses and active pharmaceutical ingredients (API), a scenario that made visible the international inequalities among countries and their varying ability to access lifesaving technologies. As part of the technology transfer strategy, Fiocruz would first receive ready-made doses and pharmaceutical ingredients as part of the learning process. This meant that until the vaccine's nationalisation, it would rely on API imported from a Chinese company designated by AstraZeneca (AZ), Wuxi Biologics. This required Bio-M to split the agreement with AZ into two contracts: the ETEC, which would cover the purchase of the API for the first vaccine doses produced by Bio-M and supplied to the government, and the actual technology transfer agreement.

You have two moments there: first the ETEC and then the technology transfer contracts. [We] chose to separate that for the sake of agility to have that start and the receipt of the API, the formulation already at Bio-M and the vaccine available as soon as possible. If we had to take care of a contract on our own, it would take much longer to close it, and we were concerned about logistics and the geopolitics surrounding the API, [...] we never know where we are in the queue. So AstraZeneca had the knife and cheese in hand. However, we also had the capital to be able to be a partner that would contribute to the success of their vaccine as well. (Interview with a Fiocruz lawyer, 2022)

AZ would initially supply the Brazilian government with enough API to manufacture 30.4 million vaccine doses. However, the pressure from the international competition for vaccines and their components was felt very quickly: in the initial AstraZeneca Brazil (AZB) proposal, made on 22 June 2020, based on the costs of the API supply (US$50 million) and the fill-finish steps to be performed by Fiocruz (US$30 million), the estimated average production cost, and the vaccine licensing fees (US$25 million), the partnership was to cost Brazil a total of US$105 million. Two days later, the company revalued the API required for 30.4 million doses, increasing its price by US$22 million, due to commitments to other countries agreed shortly prior and the restriction on API exports from some API-producing countries, thus raising the final amount to US$127 million (Brazilian Ministry of Health 2020).

The Brazilian government was awaiting the results of the Phase 3 clinical trials before committing to purchasing additional API for manufacturing the vaccines. However, after AZ informed the government that this could lead to uncertainty for the delivery of the vaccine ingredients and potentially affect its vaccination strategy, the latter decided to approve the purchase of the pharmaceutical materials necessary to produce an additional 70 million doses.[24] The health crisis was felt not only in the scarcity of health goods but also in the inflation of prices that affected national public health policies in the South. The risky investments from the South helped expand global demand, and the global sociotechnical system needed to supply these ingredients.

A second key aspect of the negotiation was ensuring that the partnership was not limited to the sale of API but also involved the complete transfer of the University of Oxford's platform. Fiocruz feared that with the technology transfer agreement taking longer to negotiate, once the company had provided the ingredients for the vaccine, it might change its mind about the transfer, something that had already happened to the Foundation in the past.[25] Thus, the Brazilian institution had to make full use of its available 'bargaining power' – in this case, the vaccine was not ready.

A third concern was the impact of intellectual property rights and the contract clauses on Bio-M's exploration of the transferred platform. This entailed verifying whether the vaccine, owned by the University of Oxford and later licensed to AZ, was not affected by patents owned by third parties that were not included in the agreement with Bio-M. Moreover, it involved negotiating with AZ the 'freedom to operate' for using the platform without incurring legal liability for this use. Patents would not be an obstacle in this case, as there were no patents granted or pending for the Oxford vaccine platform in Brazil (Fonseca et al 2023). However, it was still necessary for Bio-M to negotiate the use of all other non-patented information necessary to set up manufacturing after the transfer period, so that the company could explore the platform and even make improvements or develop new uses for it.[26] As explained by a Fiocruz lawyer when describing more broadly the negotiation of technology transfer agreements:

> Usually, these technology transfer agreements occur like this: you will learn to make the product exactly the way the guy makes it, following the whole standard; if there is any modification, it has to be approved by the partner, so you cannot change the product, you will make it exactly like he makes it and you have inherent learning by the people who work in this, they assimilate the technology, right? So you have know-how relating to the use, and you also have the IP rights associated with that technology [...] with regards to the industrial secret of the know-how, they pass it on with instructions. Finally, with regards to patents, you have to respect them, but the patents become public at some point, right? However, many things are

not even written on the patent. So, the company needs to pass on – there are techniques for writing the patent whereby you do not disclose exactly what makes up the raw material or the formulation process … a temperature range, for example … So, we have to assimilate all of this. Then we have to work on the contract, a clause that guarantees us the freedom to use it at the end, right? So, we make several types of contracts, but as a rule, we try to guarantee that after the contract expires, we have the freedom to use the technology that was transferred without paying royalties. In some cases, when a patent survives, you still have to pay a percentage of royalties for a while, but we try to guarantee in the contracts that we will have the freedom to use the technology after the contract expires. (Interview with Fiocruz lawyer, 2022)

In this sense, whereas technology transfer instruments may contribute to the implementation of national vaccination plans, their clauses might establish more or less freedom, and their negotiation has decisive implications for long-term pharmaceutical sovereignty in the South.

THE VACCINE GEOPOLITICS OF THE SOUTH AND THE LOGISTICS OF IMPORTS

In this subsection, we discuss how the process of establishing national vaccine manufacturing was affected by geopolitics at both national and global levels by looking back at two episodes involving the importing of critical ingredients and doses from India and China to support the Brazilian vaccination strategy, as well as the equally fundamental role of logistics in the process. It contributes to discussing different forms of technological globalisation at play, their role in affirming pharmaceutical sovereignty, and, finally, the nuances of globalisation promoted by Northern and Southern actors.

According to Bio-M's strategy of acquiring vaccine manufacturing capacity, the steps towards acquiring the production process started with AZ supplying the API to Bio-M, which performed the fill-finish of the vaccine doses to be supplied to the government; then, Bio-M learned and carried out the full manufacturing process, including the production of the drug substance (Fonseca et al. 2023).

Bio-M was relying on this process to start supplying the first 30 million doses manufactured with the drug substance. These components would come from the Chinese company Wuxi Biologics, one of the certified contract manufacturing organisations licensed by AZ to produce its COVID-19 vaccine components. But restrictions imposed by the Chinese government led to delays in receiving the drug substance. They affected the initial Bio-M plan to supply immunisation to the Brazilian government in December 2020, when the United Kingdom started vaccination with the same AZ shots. This led the federal government to look to import ready-to-use AstraZeneca vaccine shots. The Brazilian government looked for doses in the United States and the United Kingdom but failed. The Serum Institute of India (SII), another of AZ's certified contract manufacturing organisations, agreed to supply 2 million doses to the Brazilian government. At US$5.25 per unit, however, the price was almost twice as high as that previously agreed with AZ ($3.16 per unit) (Brazilian Ministry of Health 2020). This shows how the existence of Southern players enables pharmaceutical plans in other parts of the South, but their dynamic, too, is less about solidarity than market opportunities.

This deal was affected by the Indian government's restriction on exports of locally produced vaccines. The Brazilian Ministry of Health had been elaborating its vaccine strategy with the support of ANVISA, which had been reviewing regulatory applications to temporarily authorise the AstraZeneca and Sinovac vaccines. It ultimately issued emergency use authorisations for these two vaccines on 17 January 2021. A few days after obtaining the use authorisation, the federal government announced that it would be sending a plane to India to collect the doses from the SII. However, the Indian government discouraged the Brazilian flight, stating that it had the final say in sending the doses and would announce its decision later. Once India issued this statement, Brazil was among the first vaccine destinations, which ultimately reached the country on a flight departing from Delhi on 22 January 2021 (Banerjee 2021).

Although the Brazilian government had negotiated additional doses from the SII, by late February 2021, one month after receiving another 2 million doses, there was growing uncertainty about when the remaining doses (12 million had been negotiated) would arrive. This uncertainty was heightened by the resurgence of COVID-19 cases in India in April 2021, with the country becoming the new epicentre of the disease and interrupting the export of locally manufactured doses. In a press release, the Brazilian government affirmed its solidarity with the Indian government and anticipated that the situation would further affect its vaccination programme (Banerjee 2021). This first episode reveals how players from the South (India) have become both part of globalisation from the North and actors that enable national technological enterprises in other Southern countries. The exercise of pharmaceutical sovereignty in Brazil is affirmed by its decision to manufacture the COVID-19 vaccines locally instead of importing them. However, the process still requires importation even though it is 'temporary'. The national interests of the North and South are interrelated to globalisation processes. The paradox is that the pandemic was an opportunity to expand vaccine markets for North and South while contradictorily serving as a counter-movement to globalization through the closing of frontiers and restriction of vaccine exportation.

The second episode involved importing drug substances from Chinese companies to enable Bio-M to manufacture the first 30 million doses of the AZ vaccine. Butantan's vaccine also relied on importing drug substances from China, as it had partnered with Sinovac to locally manufacture the CoronaVac. The original timeline for receiving the API for both vaccines was affected by delays in receiving the ingredients. Bio-M and Butantan had to wade through obstacles imposed by the Chinese government to release the ingredients for export. While this decision ultimately tells of the subordination of Brazil's strategy to Chinese vaccine nationalism – that is, China reserving Chinese technologies for its vaccine strategy – there was speculation among experts and the media in Brazil about other reasons for these delays.

At around this time, President Bolsonaro was speaking out publicly against China on many occasions, suggesting its responsibility for the virus and questioning the quality of the Chinese vaccines. These statements aligned with Donald Trump's policy of denialism and confrontation with China and contrasted with the friendlier approach to India (Banerjee 2021). Bolsonaro's attitude also seemed to be informed by national politics, mainly that the governor of São Paulo, João Doria, from the right-wing party PSDB – Bolsonaro's primary political opponent at the time – had come out defending vaccines. The Butantan Institute, which belongs to the state of São Paulo, had partnered with the Chinese company Sinovac to produce the latter's vaccine, CoronaVac.

As a result, though a clear pandemic denialist and vaccine sceptic, the president ultimately supported the Bio-M vaccine strategy, mainly to capitalise on it in his struggle against his political opponent. This behaviour also had health implications by spreading doubt among Brazilian society regarding the quality of CoronaVac, the vaccine that was first made available in the country (Gramacho and Turgeon 2021), because of its Chinese origin. The politicisation of vaccines by the federal and São Paulo governments created unprecedented competition between the two public laboratories, which raised additional challenges for the stakeholders involved in the public health response to the pandemic. Conversely, frustrated by the many delays to the arrival of APIs, health expert Margareth Dalcomo – a researcher from Fiocruz who had become a media figure in the fight against misinformation during the pandemic – criticised the government's foreign policy for its inability to secure the supply of vaccines. In a television interview, the researcher stressed that all vaccines were made in China, and the ingredients for both the Fiocruz and Butantan vaccines came from that country.[27]

This second episode indicates another aspect of technological globalisation from below at play, related to the sociotechnical imaginaries built around what comes from the South (Jasanoff and Kim 2015). China had partnered with the Brazilian vaccine enterprises by taking part in development initiatives and selling drug substances, but this didn't prevent it becoming the target of attacks from President Bolsonaro, whose imaginary of China was aligned with that of US President Trump. This reveals how globalisation from below also involves dealing with powerful imaginaries that question the South based on comparisons with norms and quality standards imposed and assumptions defined by actors in the North.

According to the manager in charge of imports at Bio-M, when it came to ensuring delivery of vaccine doses and ingredients, another factor was equally as important as the political context, diplomatic negotiations, and various business dealings required. This factor was the existing infrastructure and the logistics that needed to be established to receive these technologies during a time of uncertainty and emergency, and to ensure their use without wastage. Importing the API vaccine involved organising a whole sequence of steps for the ingredients – including ingredients that required transportation at very low temperatures (between -65 and -95 degrees Celsius) – to reach Bio-M 'as fast, cheaply, and safely as possible', according to the manager of the Imports Division:

> A few months before we imported [...], we mapped the flow; we asked first the quality control departments of Bio-Manguinhos importers what would be the deadlines, the temperature, the appropriate transport equipment, what would be the best route, the companies, the reliable partners who had the know-how, expertise in transporting perishable products, vaccines. [...] We drew up contingency plans – both our experience here of importing vaccines previously and our partners' were fundamental. For example, the pharmacist of RioGaleão [the international airport Tom Jobim, in Rio de Janeiro] had the idea and convinced RioGaleão ... to rent a refrigerated container just for the API of the COVID-19 vaccine, [...] the API, because it was something new and needed to be here very fast, we adopted it, we got the Federal Revenue Service's permission, even the Federal Police's support to escort our cargo, and RioGaleão allowed our refrigerated trucks to enter the cargo terminal's yard. So, by the time our cargo left the plane, it had virtually already entered the truck. (Interview with the manager of the Imports Division at Bio-Manguinhos/Fiocruz, 2022)

All these steps were intended to ensure the integrity of the ingredients, with temperature serving as the leading indicator. In other words, as the staff in charge of imports admitted, in addition to contract negotiations, the temperature was also a significant concern within the framework of the Bio-M strategy. Logistics were required and involved a legal, technical, and political complex that could organise the flows of trade goods – in this case, vaccines and drug substances. On a material level, this also involved instruments and infrastructure such as refrigerators, flights, and telephone cameras (Quet 2018: 202). This logistics organisation is a crucial point of globalisation, enabling technological expansion and circulation.

When the plane took off, we ran there to take pictures of the temperature monitor. [...] So, the agreements were difficult; there were meetings no matter what time it was because we talked to people from China, from England, we had so many people involved in the meeting that I did not know. [...] However, when we stopped at the airport to wait for the cargo, that was that moment: 'I do not have much to do now; I just hope that everything works out when it gets here'. Then we ran to take pictures of the temperature; I sent them to the WhatsApp group where almost the whole company was. I sent the picture of the temperature, and there was that 'whew, it arrived!'. With all our effort, money, and hope, many people were waiting to get this vaccine.

CONCLUSION

This chapter contributes to debates on globalisation and its interlinkage with sovereignty through the analysis of the Brazilian COVID-19 manufacturing model based on a public-private partnership involving technology transfer. The analysis of a technological enterprise set up during a global health emergency sheds light on how globalisation has become a multidirectional phenomenon. Brazilian actors participate in at least three forms of technological globalisation.

Firstly, Brazilian vaccine initiatives participate in globalisation from above. By partnering with AstraZeneca and the United Kingdom, the Brazilian actors show their agency by coordinating the negotiation of access to Northern technology and knowledge transfer and ultimately contributing to technological development by assuming the risks of innovation. While the legal innovation and previous negotiation on TT granted BioM full access to know-how and drug ingredients, and enabled full nationalization of the Oxford vaccine, negotiations around the freedom to operate the platform after the transfer, and to the vaccines' further uses, constituted a major challenge. In this sense, Brazilian actors may engage and contribute with globalisation from above to address their public health interests, but this engagement may condition the fulfilment of pharmaceutical sovereignty purposes in the future.

However, when negotiating contractual terms defining the freedom of use, Brazilian institutions are in a less favourable position than many Northern players in the current global pharmaceutical system. Brazil was able to engage and reaffirm the global IP rules by not choosing the 'confrontational approach' employed in the copy of HIV drugs (the use of patent flexibility for reverse engineering), not considered appropriate for the local production of a biological product, particularly a new vaccine platform, which requires access to knowledge not present in patents.

Secondly, Brazilian actors also participate in technological globalisation alongside Indian and Chinese pharmaceutical actors. These countries have become inescapable global suppliers of vaccine shots and ingredients, and their presence in the global pharmaceutical value chain reframes relationships of power with the North. In the scenario of international inequality, the emergence of Southern alternatives enables technological enterprises in the South through opportunities for new forms of commercial partnerships and technological agreements and the expansion of Southern technological markets. However, these relationships are unstable, particularly with the yet-to-be-understood impact of influential emerging Asian players and the varying sociotechnical imaginaries of the Chinese and Indian presence in Brazil and other Southern countries, with potential effects on sovereignty.

Thirdly, Brazilian actors are involved in the globalisation of their technologies. The primary purpose of the public biotechnology manufacturing model in Brazil is to affirm sovereignty and ensure the implementation of national health policies for all Brazilians in an unequal country. Although they have less global reach than Chinese and Indian industries, Brazilian public laboratories are essential global suppliers of yellow fever and influenza vaccines, with the COVID-19 pandemic offering new impetus for the expansion of this model. Moreover, the advances of indigenous vaccine development in Brazil during COVID-19, though not timely to address the pandemic, indicate both the existence of R&D capacity and bottlenecks in the national innovation system. Although Brazilian institutions have been shifting their discourses to embrace the approach of nationally developed platforms, these require long-term investments. Freedom to operate clauses in technology transfer agreements should be seriously considered in national and regional vaccine manufacturing plans.

Finally, this chapter can only be finished by considering the agency of the SARS-CoV-2 virus that obliges us to reconsider the globalisation phenomenon. The virus behind the COVID-19 pandemic exposed the contradictory scenario of pharmaceutical market expansion and the damaging effects of globalisation forms based on dependency, either on the North or the emerging South. The investigation of the Brazilian vaccine initiative indicates how, in a scenario of uneven pharmaceutical geography, access to vaccines in the South is enabled by the creative assemblage of the national and the global (Ong 2004). Sovereignty related to pharmaceutical manufacturing plays a role in the equitable distribution of health innovations required to face a global health crisis. However, this bioproduction-related sovereignty is not straightforward to affirm, requiring complex legal instruments to negotiate (van Wichelen 2023) or long-term investment in local innovation. A better global scenario would be supporting countries and regions through relationships of solidarity in the promotion of their capacity to equitably address their populations' needs related to vaccines and other public health goods.

ENDNOTES

1 By January 2022, four vaccines had been mobilised as part of the vaccine strategy: 115.6 million AstraZeneca doses (36.9%); 103 million Pfizer doses (33.2%); 8.2 million Butantan/Sinovac doses (26.9%); and 4.8 million Janssen doses (1.5%).

2 'Classic' vaccine platforms are those based on already known technologies, such as virus-based or protein-based vaccines, which represent the majority of vaccines licensed for human use. Examples include inactivated-virus vaccines, such as polio vaccines, or vaccines based on a protein purified from the virus, such as the seasonal influenza vaccine. Next-generation vaccines can be developed based on sequence information alone and, therefore, do not require culturing the virus, making these platforms more adaptable and speeding up vaccine development. These vaccines include platforms based on viral vectors, DNA, RNA, and antigen-presenting cells (Riel and Wit 2020).

3 The Russian vaccine Sputnik V, developed by the Gamaleya National Center of Epidemiology and Microbiology in Moscow, was the first registered vaccine against COVID-19. It is a non-replicating viral vector-based vaccine, similar to the Oxford–

AstraZeneca and Janssen vaccines. Gamaleya made 54 purchase agreements with 52 countries, with 98.79% of the doses committed to lower- and upper-middle-income countries (Ruiz and Tadevosyan 2022). India's Serum Institute, the world's largest vaccine producer, was chosen as the COVAX Facility's leading vaccine supplier. The Indian government also invested in an indigenous vaccine, Covaxin, developed by Bharat Biotech and ICMR, which obtained WHO emergency authorisation. However, the latter was contested at both national and global levels; the WHO suspended the emergency authorisation after inspections of Bharat Biotech's plant in April 2022.

4 Members of civil society have particularly championed the domestic production of vaccines by developing countries, as shown by a Médecins Sans Frontières report on the capacity of laboratories in developing countries to produce messenger-RNA vaccines (Médecins et al. Campaign 2021). Local vaccine production has been proposed before, particularly in a WHO report (World Health Organisation 2011).

5 Data were collected during fieldwork as part of the 'Uneven Geographies of Vaccine R&D'' project funded by the German Research Foundation. The data collection was equally supported by the Institut Francilien Recherche Innovation Société (IFRIS), as part of the author's postdoctoral fellowship.

6 The Commission was installed on 27 March 2021 for 90 days (renewed for the same period) at the Federal Senate to investigate 'actions and omissions from the Federal Government in confronting the COVID-19 pandemic in Brazil'. Brazilian Federal Senate. (2021). *CPI da pandemia é prorrogada por mais 90 dias* [Pandemic CPI extended for another 90 days]. Senado Federal, July 14 https://www12.senado. leg.br/noticias/materias/2021/07/14/cpi-da-pandemia-e-prorrogada-por-mais-90-dias.

7 In addition to the public market (PNI), a private market has emerged in Brazil to provide access to modern vaccines that are unavailable in the public health sector. While it creates inequality in the country, this private market does provide a gateway for more technologically advanced vaccines to enter the public market (Gadelha et al. 2020).

8 See World Health Organization. *Prequalified vaccines:* https://extranet.who.int/ prequal/vaccines/prequalified-vaccines

9 Oswaldo Cruz Foundation. *Units and offices.* Fiocruz: https://portal.fiocruz.br/ en/units-and-offices

10 The precursor of Fiocruz and Bio-M was the Oswaldo Cruz Institute (IOC), now the Foundation's central research unit. The IOC was the original institution created in 1900 to produce vaccines and serum against the major infectious diseases affecting the country's socioeconomic development (bubonic plague, smallpox, and yellow fever) (Benchimol 2017; Löwy 2006). Fiocruz was established in the 1970s as a network of institutions, with biological production activities to be handled by Bio-M, research by IOC, and units in other states and education by the School of Public Health.

11 Bio-Ms, along with the other Fiocruz research and production units, were in a precarious position at that time. By creating Fiocruz in 1970, the government sought to leverage national science and technology for its health programmes, as the country had done earlier in the twentieth century. However, Fiocruz's research and production facilities lacked public investment in the early decades (Stepan 1976). The Butantan Institute also suffered from a similar situation of precarious infrastructure and decay due to a lack of investment from the public authorities.

12 These institutions were the Mérieux Institute, the Biken Institute, and the Japan Poliomyelitis Research Institute.

13 The PAHO Revolving Fund and UNICEF are the two largest procurers of vaccines for public-sector immunisation programmes (Milstien et al. 2007).

14 The productive process of a vaccine involves five phases: the production of the vaccine concentrate, the vaccine formulation, fill-finish, quality control of the product, and productive process analysis (Silva 2005).

15 Until then, the most expensive vaccine supplied to the government had been the measles vaccine, at US$0.30 per dose, whereas the vaccine against Hib was commercialised at US$2.50 per dose. The agreement covered the provision of about 60 million doses, meaning that the total revenue from the sale of this vaccine to the Ministry amounted to US$150 million (Ponte 2007).

16 List of WHO Prequalified yellow fever vaccines: https://extranet.who.int/pqweb/vaccines/prequalified-vaccines?field_vaccines_effective_date%5Bdate%5D=&field_vaccines_effective_date_1%5Bdate%5D=&field_vaccines_type%5B0%5D=Yellow%20Fever&field_vaccines_name=&search_api_views_fulltext=&field_vaccines_number_of_doses=&page=1.

17 Far-Manguinhos/Fiocruz has a particular role in making available antiretroviral medicines in Brazil. It adopted a policy of copying these medicines through reverse engineering in the 1990s. With the recognition of pharmaceutical patents after the signature of the TRIPS Agreement, the copy strategy was prohibited for new and patented medicines. The use of patent flexibilities – including compulsory licensing and patent opposition – was part of formulating the pharmaceutical policy. Far-Manguinhos filed patent opposition to some HIV medicines, and the menace of compulsory licensing was used to negotiate price reductions for these medicines.

18 Industrial policy was certainly affected by a political crisis that followed President Dilma Rousseff's impeachment (2010–2016) and her replacement in 2016 by Michel Temer, from MDB, with a more right-wing approach. Despite this period, especially the government of Jair Bolsonaro (2018–2022), the PDPs were not suppressed.

Since the election of Luiz Inacio Lula da Silva in 2023, the politics of the 'healthcare-economic-industrial complex' has been re-established and even expanded.

19 In the vaccine sector, this has included partnerships with some players that concentrate global revenue from vaccines (GSK, Merck, and Sanofi Pasteur) (Gadelha et al. 2020).

20 The government evaluated proposals from the following companies: Janssen, Sanofi, Moderna, Pfizer, University of Oxford-AZ, Inovio, Sinovac, and CanSino.

21 Jornal da USP. (2020, June 11). *Instituto Butantan faz parceria com empresa chinesa para testar vacina contra SARS-CoV-2* [Instituto Butantan partners with Chinese company to test vaccine against SARS-CoV-2]. University of São Paulo. https://jornal.usp.br/ciencias/instituto-butantan-faz-parceria-com-empresa-chinesa-para-testar-vacina-contra-sarscov2/

22 The Ministry of Health enlisted Bio-M to find solutions to enhance access to vaccines, which resulted in meetings with AZ to explore the possibility of national production of their vaccine. The ambassador of the United Kingdom also attended a technical meeting in June, which led to the Embassy's formal request to the Brazilian Ministry of Health to establish an agreement with AZ (Brazilian Ministry of Health 2020).

23 As the Bio-M representative explained, the production processes for the Oxford COVID-19 vaccine and alfapoetina mobilised similar platforms, as both involved the use of cell culture in bioreactors. However, the vaccine's manufacturing process had specific characteristics and employed more recent technology.

24 As mentioned in the *Nota Informativa N.1/2020-CGPCLIN/DECIT/SCTIE/MS* of the Ministry of Health, dated 26 June 2020, sent in a letter to the British ambassador in Brazil, AZ warned that waiting for the results of the Phase 3 trials would result in this additional API only being available in the third trimester of 2021 (Brazilian Ministry of Health 2020).

25 This happened on at least two occasions: in the 1980s, when the Mérieux Institute gave up on transferring the vaccine for meningitis, and in the 2000s, when Applied Biosystems gave up on transferring an HIV viral-load platform one year after reaching an agreement with Bio-M to this end (Kameda 2021; Pontes 2007).

26 It was also important to consider further developments led by AZ that could improve the vaccine and ensure that Bio-M would be able to incorporate these as well (interview with Fiocruz lawyer, 2022).

27 GNT, interview on 18 December 2020, https://www.facebook.com/watch/?v=209816200643510.

REFERENCES

ANVISA, Agência Nacional de Vigilância Sanitária, 'Anvisa aprova por unanimidade uso emergencial das vacinas', *Agência Nacional de Vigilância Sanitária*, 17 January 2021 https://www.gov.br/anvisa/pt-br/assuntos/noticias-anvisa/2021/anvisa-aprova-por-unanimidade-uso-emergencial-das-vacinas.

Banerjee, D., 'From Internationalism to Nationalism: A New Vaccine Apartheid', *Comparative Studies of South Asia, Africa and the Middle East*, 41.3 (2021): 312–17 https://doi.org/10.1215/1089201X-9407806.

Benchimol, J., 'Yellow Fever Vaccine in Brazil: Fighting a Tropical Scourge, Modernising the Nation', in Christine Holmberg, Stuart Blume, and Paul Greenough, eds., *The Politics of Vaccination: A Global History* (Manchester University Press, 2017), pp. 176–94 https://doi.org/10.7228/manchester/9781526110886.003.0008.

Biehl, J. G., *Will to Live: AIDS Therapies and the Politics of Survival* (Princeton University Press, 2007).

Birch, K., ed., *Assetization: Turning Things into Assets in Technoscientific Capitalism* (MIT Press, 2020).

Blume, S., *Immunisation: How Vaccines Became Controversial* (Reaktion Books, 2017) https://press.uchicago.edu/ucp/books/book/distributed/I/bo27430344.html.

Blume, S., and B. Baylac-Paouly, *Immunisation and States: The Politics of Making Vaccines* (Routledge, 2021) https://doi.org/10.4324/9781003130345.

Blume, S., and M. Zanders, 'Vaccine Independence, Local Competences and Globalisation: Lessons from the History of Pertussis Vaccines', *Social Science & Medicine*, 63.7 (2006): 1825–35 https://doi.org/10.1016/j.socscimed.2006.04.014.

Brazilian Ministry of Health, *Nota Técnica Conjunta n° 25/2020 – CGPCLIN/DECIT/SCTIE/MS (0016051533)* (Brasília Ministry of Health, 2020).

Buss, P. M., Temporão, J. G., & Carvalheiro, J. R. *Vacinas, soros e imunizações no Brasil.* (Editora Fiocruz, 2005). https://doi.org/10.7476/9788575416068

Cassier, M., and M. Correa, *Health Innovation and Social Justice in Brazil* (Springer International Publishing, 2018) https://doi.org/10.1007/978-3-319-93154-2.

Chang, H-J., *Kicking Away the Ladder: Development Strategy in Historical Perspective* (Anthem Press, 2002).

Chatterjee, N., Z. Mahmood, and E. Marcussen, 'Politics of Vaccine Nationalism in India: Global and Domestic Implications', *Forum for Development Studies*, 48.2 (2021): 357–69 https://doi.org/10.1080/08039410.2021.1918238.

Chaves, G. C., L. Hasenclever, and M. A. Oliveira, 'Conexões Entre as Políticas de Desenvolvimento Industrial no Setor Farmacêutico e a Política de Saúde no Brasil: Um Percurso da Década de 1930 a 2000', in *Desafios de Operação e Desenvolvimento do Complexo Industrial da Saúde* (Rio de Janeiro, e-papers, 2016).

Clemens, S., *História de Uma Vacina: O Relato da Cientista Brasileira Que Liderou os Testes da Vacina Oxford/AstraZeneca no País* (História Real, 2021).

Edquist, C., & Zabala-Iturriagagoitia, J. M. Public procurement for innovation as mission-oriented innovation policy. *Research Policy, 41*(10) (2012) 1757–1769. https://doi.org/10.1016/j.respol.2012.04.022

Fearnley, L. Viral sovereignty or sequence etiquette? Asian science, open data, and knowledge control in global virus surveillance. *East Asian Science, Technology and Society: An International Journal, 14*(3) (2020) 479–505. https://doi.org/10.1215/18752160-8698019

Flynn, M., 'Brazilian Pharmaceutical Diplomacy: Social Democratic Principles Versus Soft Power Interests', *International Journal of Health Services*, 43.1 (2013): 67–89 https://doi.org/10.2190/HS.43.1.f.

Fonseca, E. M. da, and others, 'Vaccine Technology Transfer in a Global Health Crisis: Actors, Capabilities, and Institutions', *Research Policy*, 52.4 (2023): 104739 https://doi.org/10.1016/j.respol.2023.104739.

Gadelha, C. A. G., and others, 'Acesso a Vacinas no Brasil no Contexto da Dinâmica Global do Complexo Econômico-Industrial da Saúde', *Cadernos de Saúde Pública*, 36.Suppl. 2 (2020): e00154519 https://doi.org/10.1590/0102-311x00154519.

Garrison, C., 'How the "Oxford" Covid-19 Vaccine Became the "AstraZeneca" Covid-19 Vaccine', *Medicines Law & Policy* (October 5, 2020) https://medicineslawandpolicy.org/2020/10/how-the-oxford-covid-19-vaccine-became-the-astrazeneca-covid-19-vaccine/.

Global Health Centre, *COVID-19 Vaccine Manufacturing* (Graduate Institute of International and Development Studies, 2021) https://www.knowledgeportalia.org/covid19-vaccine-manufacturing.

Gramacho, W. G., and M. Turgeon, 'When Politics Collides with Public Health: COVID-19 Vaccine Country of Origin and Vaccination Acceptance in Brazil', *Vaccine*, 39.19 (2021): 2608–12 https://doi.org/10.1016/j.vaccine.2021.03.080.

Hasenclever, L., Paranhos, J., Chaves, G. C., & Oliveira, M. A. *Vulnerabilidades do complexo industrial da saúde: Reflexos das políticas industrial e tecnológica na produção local e assistência farmacêutica*. Editora E-Papers. (2018) https://www.e-papers.com.br/produto/vulnerabilidades_do_complexo_industrial_da_saude/?v=82a9e4d26595

Hayden, C., 'A Generic Solution?: Pharmaceuticals and the Politics of the Similar in Mexico', *Current Anthropology*, 48.4 (2007): 475–95 https://doi.org/10.1086/518301.

Instituto Butantan, 'Retrospectiva 2021: Segundo Ano da Pandemia é Marcado Pelo Avanço da Vacinação Contra Covid-19 no Brasil', *Instituto Butantan*, 31 December 2021 https://butantan.gov.br/noticias/retrospectiva-2021-segundo-ano-da-pandemia-e-marcado-pelo-avanco-da-vacinacao-contra-covid-19-no-brasil.

Jasanoff, S., and S-H Kim, eds., *Dreamscapes of Modernity: Sociotechnical Imaginaries and the Fabrication of Power* (University of Chicago Press, 2015).

Jensen, N., A. Barry, and Ann H. Kelly, 'More-than-National and Less-than-Global: The Biochemical Infrastructure of Vaccine Manufacturing', *Economy and Society*, 0.0 (2022): 1–28 https://doi.org/10.1080/03085147.2022.2087899.

Kameda, K., 'Molecular Sovereignty: Building a Blood Screening Test for the Brazilian Nation', *Medicine Anthropology Theory*, 8.2 (2021): 1–25 https://doi.org/10.17157/mat.8.2.5122.

Kameda, K., A. H. Kelly, J. Lezaun, and Ilana Löwy, 'Imperfect Diagnosis: The Truncated Legacies of Zika Testing', *Social Studies of Science* (2021): 683–706 https://doi.org/10.1177/03063127211035492.

Kelly, A. H., and others, 'Uncertainty in Times of Medical Emergency: Knowledge Gaps and Structural Ignorance During the Brazilian Zika Crisis', *Social Science & Medicine*, 246 (2020): 112787 https://doi.org/10.1016/j.socscimed.2020.112787.

Lezaun, J., and C. M. Montgomery, 'The Pharmaceutical Commons: Sharing and Exclusion in Global Health Drug Development', *Science, Technology & Human Values*, 40.1 (2015): 3–29 https://doi.org/10.1177/0162243914542349.

Löwy, I., *Vírus, Mosquitos e Modernidade: A Febre Amarela no Brasil Entre Ciência e Política* (Editora Fiocruz, 2006).

Loyola, M. A., 'Medicamentos e Saúde Pública em Tempos de AIDS: Metamorfoses de Uma Política Dependente', *Ciência & Saúde Coletiva*, 13.Suppl. (2008) http://www.scielo.br/pdf/csc/v13s0/a27v13s0.pdf.

Lurie, N., M. Saville, R. Hatchett, and J. Halton, 'Developing COVID-19 Vaccines at Pandemic Speed', *New England Journal of Medicine*, 382.21 (2020): 1969–73 https://doi.org/10.1056/NEJMp2005630.

Médecins Sans Frontières. *Local diagnostics to meet local health needs: Médecins Sans Frontières Access Campaign.* (2021) https://msfaccess.org/improve-local-production-diagnostics

Medeiros, M. Z., Soares, P. F., Fialho, B. C., Gauss, L., Piran, F. S., & Lacerda, D. P. Vaccine innovation model: A technology transfer perspective in pandemic contexts. *Vaccine*, 40(33) (2022) 4748–4763. https://doi.org/10.1016/j.vaccine.2022.06.054

Medina, E., I. Marques, and C. Holmes, eds., *Beyond Imported Magic: Essays on Science, Technology, and Society in Latin America* (MIT Press, 2014).

Milstien, J. B., Gaulé, P., & Kaddar, M. Access to vaccine technologies in developing countries: Brazil and India. *Vaccine, 25*(44) (2007): 7610–7619. https://doi.org/10.1016/j.vaccine.2007.09.007

Ortega, F., and M. Orsini, 'Governing COVID-19 Without Government in Brazil: Ignorance, Neoliberal Authoritarianism, and the Collapse of Public Health Leadership', *Global Public Health*, 15.9 (2020): 1257–77 https://doi.org/10.1080/17441692.2020.1795223.

Paremoer, L., and A. Pollock, '"A Passion to Change the Landscape and Drive a Renaissance": The mRNA Hub at Afrigen as Decolonial Aspiration', *Frontiers in Public Health*, 10 (2022): 1065993 https://doi.org/10.3389/fpubh.2022.1065993.

Peixoto, V. D. M., Leal, J., and L. M. Marques, 'The Impact of Bolsonarismo on COVID-19 Vaccination Coverage in Brazilian Municipalities', *Saúde Em Debate*, 47.139 (2023): 806–17 https://doi.org/10.1590/0103-1104202313906.

Pollock, A., *Synthesizing Hope: Matter, Knowledge, and Place in South African Drug Discovery* (University of Chicago Press, 2019).

Ponte, C. F., 'Bio-Manguinhos 30 anos: A trajetória de uma instituição pública de ciência e tecnologia', *Cadernos de História da Ciência*, 3.1 (2007): 35–138 https://doi.org/10.47692/cadhistcienc.2007.v3.35696.

Quet, M., *Illicit Medicines in the Global South: Public Health Access and Pharmaceutical Regulation* (Routledge, 2021).

Quet, M. *Impostures pharmaceutiques: Médicaments illicites et luttes pour l'accès à la santé.* Les Empêcheurs de penser en rond. (Paris: La Découverte, 2018) https://www.cairn.info/impostures-pharmaceutiques--9782359251418.htm

Riel, D. van, & de Wit, E. 'Next-generation vaccine platforms for COVID-19'. *Nature Materials, 19*(8) (2020) 810–812. https://doi.org/10.1038/s41563-020-0746-0

Ruiz, A. A., and G. Tadevosyan, 'Russian COVID-19 Vaccines', Knowledge Portalia (2022) https://www.knowledgeportalia.org/russian-covid-19-vaccines.

Silva, F. I. da. *O planejamento e controle de produção para uma fábrica de vacinas.* (2005) https://www.arca.fiocruz.br/handle/icict/33602

Stepan, N. *Beginnings of Brazilian science: Oswaldo Cruz, medical research and policy, 1890–1920.* (Science History Publications, 1976)

STOPAIDS, 'Access Denied: The Role of Trade Secrets in Preventing Global Equitable Access to COVID-19 Tools', STOPAIDS Blog (2023) https://stopaids.org.uk/access-denied-report-series-launch.

Sunder Rajan, K., *Pharmocracy: Value, Politics, & Knowledge in Global Biomedicine* (Duke University Press, 2017).

van Wichelen, S., 'After Biosovereignty: The Material Transfer Agreement as Technology of Relations', *Social Studies of Science*, 53.4 (2023): 599–621 https://doi.org/10.1177/03063127231177455.

Wang, X., 'Chinese COVID-19 Vaccines', Knowledge Portal (2022) https://www.knowledgeportalia.org/chinese-vaccines.

WHO (World Health Organisation), *Global Vaccine Market Report 2022: A Shared Understanding for Equitable Access to Vaccines* (World Health Organisation, 2023) https://iris.who.int/bitstream/handle/10665/367213/9789240062726.

WHO (World Health Organization). *Local production for access to medical products: Developing a framework to improve public health.* (World Health Organization, 2011)

Zhang, M., and L. Bjerke. 'Antibiotics 'Dumped': Negotiating Pharmaceutical Identities, Properties, and Interests in China–India Trade Disputes.' *Medical Anthropology Quarterly* 37(2) (2023).: 148–163. https://doi.org/10.1111/maq.12757.

6

A HUMAN DRUG AMID ANIMAL DISEASES: THE ECOLOGY OF GLOBALISED HEPARIN

Thibaut Serviant-Fine

HISTORIAN DOMINICK LACAPRA (2023) HAS STATED THAT 'AS A RULE, THE problem of globalisation is confined to humans', and this presupposition has kept research at the intersection between globalisation studies and animal studies scarce. Yet, an emerging body of scholarship has been increasingly discussing the status and significance of animal life in globalisation dynamics, for example regarding the commodification processes operating in global wildlife or pet markets (Collard 2020; Haraway 2008), the impact of the globalisation of law upon animal welfare and health (Blattner 2019; Park and Singer 2012; Sykes 2016), and the use of animal materials in the fabrication of health commodities intended for international markets (Chee 2021; Gameiro and Quet 2023). In addition, the vast literature guided by the 'One Health' approach has explored the multiple ecological entanglements between human and non-human animal health, the most prominent of which consists of the spread of non-human zoonoses to humans (Brown and Nading 2019; Craddock and Hinchliffe 2015; Kelly et al 2018) and of the impact of such phenomena upon the structuration of global health organisations (Fearnley 2020; Kelly et al 2020; Lakoff 2008). Building on these contributions, the following chapter intends to delineate the entanglements of three features of contemporary globalisation: the worldwide extension of raw material procurement practices through increasingly long logistical chains, the geopolitical and economic readjustments between countries or geographic areas associated with the shifting meaning of expressions such as 'Global North' and 'Global South', and the industrial transformation of animal matter into key raw materials. In this effort, our guiding interrogation is about the role of animals in the fabrication of globalisation processes: to what extent do animal populations, especially through epidemic episodes, contribute to shaping the practices of procurement and, in turn, the relations of dependence and domination between countries labelled as belonging to the Global South or the Global North? In order to answer this question, the chapter discusses the evolution of global procurement practices pertaining to heparin, an anticoagulant drug, for its production.

The world's consumption of heparin requires vast quantities of animal matter for its manufacturing – two-thirds of which originate from the bodies of Chinese non-human animals, a factor that is crucial to understanding the global pharmaceutical industry's growing dependence on factory farming through transnational supply chains. In spite of their predominant role, the importance of animals and animal matter in this globalisation phenomenon is barely acknowledged. In global trade statistics, they exist only as pharmaceutical components, and drug manufacturers consider them as inputs. There are, however, moments when this importance gets the attention it deserves: when animals fall sick. The biomedical operations that support human health using heparin increasingly rely on the industrialisation of animal life, which has led to production on unprecedented scales, and depends upon complex and highly international sourcing operations. The growing exploitation of non-human animals for a global market has heightened the risk of emerging diseases and paradoxically increased the vulnerability of the heparin supply chain. These animal diseases have had major consequences on the global heparin supply in three recent successive episodes, which inform the chronological structure of this chapter. The first episode was the epidemic of bovine spongiform encephalopathy, or mad cow disease, in the 1990s. The second was the scandal of adulterated heparin originating from China, in 2007–2008. The last episode pertains to the global effects of the African swine fever pandemic, which have been growing since 2018. Each episode changed the patterns of heparin production and circulation; it impacted as well as it reflected the structuration of global pharmaceutical markets. The first occurred at a time when globalisation was triumphant and long-distance outsourcing relatively unquestioned in dominant discourses. With the 2008 scandal, doubts emerged regarding pharmaceutical markets, which appeared to be globalising rapidly and perhaps uncontrollably. Finally, the third episode saw the mainstream consolidation of more direct criticism of global pharmaceutical flows, amid shortages and supply chain disruptions.

Animals are not the only actors in this story: the Chinese pharmaceutical and agro-industrial companies have developed greatly over the last 30 years and have now become the main suppliers of active pharmaceutical ingredients worldwide, from multinational companies in the North to Indian generic companies, and consolidated large companies innovating with new finished drugs (Zhang and Bjerke 2023). The country's pig farmers, once food producers, have become key suppliers of the global pharmaceutical industry, competing with the animal industries of Iowa, in the USA, and Brittany, in France. Once considered subaltern actors, these companies have become major players in the global heparin industry. Highlighting the Chinese industry's position as the hub of polycentric flows helps us nuance the conventional representation of South-to-North supply circuits and demonstrates how technological and regulatory change on a global scale can also be triggered by formerly dominated actors. Through this perspective, the ecological interconnections required by heparin production bring attention to the ways in which the material products that ensure human health, their sourcing, and the flaws related to their procurement influence the globalisation of pharmaceutical markets and alter conceptions of North and South as dominant/dominated or emerged/emerging entities.

This chapter results from research conducted from 2021 to 2023 within the Anipharm project,[1] which analysed the pharmaceutical uses of animal life. It relies upon a systematic analysis of the grey documentation pertaining to global heparin markets, as well as a collection of media reports documenting the impact of zoonotic crises upon heparin procurement and production. After a section introducing heparin as a global medicine, each of the three next sections discusses one epidemic episode and analyses the articulations it revealed between animal populations, procurement chains, and geopolitical power relations. Exploring the dependence of the heparin supply chain on industrialised animals will eventually raise the question of the conditions needed to maintain the supply of and access to this drug for a large population, and the environmental sustainability of its mode of production.

HEPARIN: ANIMAL MATTER AS A GLOBAL MEDICINE

Historically, non-human animal matter has been widely processed to produce therapeutic products, across different geographical and cultural contexts. But acknowledgement of the fact that animals are still ubiquitous in human medicine today is not a given. The use of laboratory animals has long been publicised, which has led to many changes. However, the continued use of animals as drug production material is less recognised. The discovery of this persisting presence often prompts a puzzled response in Western biomedical settings. The use of animals in medicine is usually seen either as a relic of the past or as the preserve of so-called traditional medicine. Over the last century, the advancement of synthetic chemistry and subsequently biotechnology gradually reduced the use of animals as a primary source of matter for drug production, and more generally decreased dependence on harvesting natural resources for healing (Schwerin et al 2013). In parallel, rising concerns about animal welfare have also fuelled efforts to minimise animal use. However, in spite of these developments, animal use in human medicine has continued to grow. The linear narrative of therapeutic progress moving away from old remedies must be nuanced in order to account for the hidden persistence of animal matter in biomedicine. Heparin is a particularly apt example of this persistence, as its production did not follow this downward trend but quite the contrary.

Starting in the nineteenth century, the advent of new chemical tools gradually led to the analysis and processing of natural resources, driven by a quest for physiologically active materials, purified extracts, or isolated substances. At the turn of the twentieth century, with the emergence of new biochemical methods, many animal extracts were being tested for their physiological activity.

These extracts often originated from slaughterhouses, at a time when the uses of animals for human nutrition were starting to be deeply transformed by the rise of the industrial food system (Lee 2008; Specht 2020). Readily available animal matter from slaughterhouses, combined with improved technical methods for extraction and standardisation, provided experimental material for the advancement of physiology and endocrinology. This led to the discovery of hormones, the therapeutic production of which relied on the interconnection between growing pharmaceutical companies and food industries (Clarke 1995; Oudshoorn 1994). The most well-known and impactful discovery in this field was insulin, in the early 1920s, but heparin was also one such substance extracted from animals. First identified in the 1910s in dogs' livers, heparin consists of a family of very large molecules of close but variable compositions. As the first preparations were quite toxic, heparin only began to be used reliably in clinical settings after its preparation was improved, in the 1930s in Toronto, by the research group that had first prepared and used insulin. It was developed into an essential anticoagulant drug to prevent and treat embolism; prevent blood clotting in surgery, blood transfusion, and dialysis; and for other specific indications. Heparin is also used as a reagent in test tubes for a range of laboratory analyses of blood to prevent the clotting of samples, or in heparin-coated catheters, aptly illustrating how animal matter may be hidden deep within routine and essential biomedical infrastructure.

At present, heparin-based drugs are mostly manufactured using byproducts of industrial pig farming, and beef production to a lesser extent. While close to two-thirds of the global heparin supply comes from China (Agence Nationale de Sécurité du Médicament 2014: 2), it is also produced in other countries, such as the USA, Brazil, Argentina, India, France, Spain, Italy, and Germany. The distribution of the contemporary geographies of heparin production and consumption between the Global North and the Global South has become increasingly complex: a French industrial site may process raw heparin sourced either locally, from China, or from the USA to make an intermediary product, which it will then send to Singapore for the preparation of the finished drug. Likewise, raw heparin extracted in Iowa may be sent to the Netherlands for purification, then to another French factory for processing into another heparin-based drug, and

Chinese raw heparin is supplied to Indian companies that manufacture generic heparin for Asian and African markets.

Over the course of the twentieth century, while the use of animal resources to prepare drugs decreased, heparin followed the opposite trend and was increasingly consumed, mainly due to two factors. First, technological innovations relating to heparin expanded its usability and applications, especially in outpatient treatment. In particular, the development in the 1980s of low-molecular-weight heparins (LMWH), which weigh fractions of the larger unfractionated heparin molecules, reduced side effects and the need for monitoring of potential sudden and dire side effects such as heparin-induced thrombocytopenia. Second, the demand for heparin exploded. In high-income countries, the aging population was increasingly subject to medical conditions requiring anticoagulant drugs, while low- to middle-income countries also started adopting heparin in medical practice. However, as heparin consumption increased, a parallel unrelated development came to affect its distribution, namely the advent of mad cow disease.

EPISODE 1: MAD COWS IN THE NORTH AND THE TURN TO CHINA

In the mid-1980s, the epidemic of bovine spongiform encephalopathy (BSE), or mad cow disease, emerged in the United Kingdom. After characterising the disease, pathologists soon noticed that the brain tissue of affected cattle presented similar lesions to those observed in sheep affected by scrapie. The cattle disease was rapidly associated with animals that had consumed meat-and-bone feed, a foodstuff made of processed offal from a variety of animals. This type of feed was promptly banned, but cases continued to soar and peaked in 1992–1993. This generated widespread fear that the disease could be transmissible to humans, even though no such cases had occurred with scrapie. Based on its histology, it seemed that BSE could be similar to Creutzfeldt-Jakob disease (CJD), a rare neurological disorder; a monitoring unit was promptly established. In 1995, a variant of the disease, shortened as vCJD, was identified when individuals younger than those usually affected by CJD started to show symptoms: vCJD in humans was associated with eating BSE-infected meat, which the UK government officially recognised just a year later, in 1996. The suspicion that the use of beef materials to make medical products could cause the transmission of vCJD to humans led to the precautionary ban of these materials for heparin production in the late 1990s, in the USA and Europe.

While beef lungs were used until the 1950s, they were not ideal material, as the processes employed to treat them could degrade heparin and their pungent rotting raised issues with local communities. Intestinal mucosa from cows or pigs, a readily available byproduct of sausage casing manufacturing, were preferred by the industry. Paying attention to the concrete matter required in production processes allows us to ground the study of pharmaceutical flows, which are often approached through the prism of their economic dimensions. In the mid-twentieth century, a truckload of 40,000 pounds of pig mucosa produced five pounds of heparin (0.0125%) (Barrowcliffe 2012; Coyne 1981). These quantities show how closely interlinked the food and pharmaceutical industries were surrounding these specific processes, as the vast amounts of materials required meant that the slaughtering facilities and processing plants needed to be in close proximity. Heparin thus strongly maintained and even extended this connection, as its production necessarily originated in countries where animal materials were readily available (United States, France, Canada, Germany, Netherlands, Denmark, etc.).

In the 1970s, sourcing became more diversified: beyond local sources, some companies started to look for countries with large pig herds, such as China (still by far the main producer of pork worldwide). When BSE struck, pig intestines were already favoured, but with the ban on beef intestines as another available source, the pharmaceutical companies concerned quickly moved to secure their sourcing of pig material. This significantly accelerated the general turn towards China and expanded the supply chain to secure the necessary materials as consumption grew. At the time, the potential risks associated with this expanding supply chain and the reduction of animal sources from two main species to one were not clearly described or questioned, as though material flows would naturally keep up with the need to source elsewhere.

The increase in heparin consumption had given rise to staggering demand for raw materials. An investigation for a French consumer magazine noted that half of global heparin production came from Chinese pigs, with an estimated 500 million intestines per year needed (Browaeys 2005). A 2014 report by the French national agency for drug safety quoted the same figure and estimated that 55–60% of these pigs came from China (Agence Nationale de Sécurité du Médicament 2014: 2). In a 2016 interview, Ruixin Miao, head of the heparin-trading company Nanjing Kaiyang Biotech, estimated that 'owing to productivity improvements, it now takes about 1,500 intestines to produce 1 kg of heparin' (Tremblay 2016). He estimated that 300 million pigs in China could produce 17.6 million megas (a measurement unit) of heparin, and that global demand amounted to 28 million megas. According to his estimate, the number of pigs required to meet global demand stood at 477 million, close to the figures quoted above. Comparing these numbers to the global availability of pork material gives a sense of the level of dependency on industrial farming for the production of this essential drug: while 1.3–1.5 billion pigs have been slaughtered every year since 2010, this number was closer to 1 billion around the year 2000, and if we go back to the 1960s, the 500 million intestines required would have accounted for the entire global pig population.

EPISODE 2: DYING PIGS AND DEADLY FRAUD

The second episode occurred against the backdrop of growing heparin consumption and the broader global shift toward China as a key supplier. Amid this trend, a major scandal emerged around the adulteration of heparin sources. This involved the inclusion of oversulphated chondroitin sulphate, a substance hard to differentiate from heparin. The likely motive for this adulteration was the sudden scarcity of pigs (and their materials) due to a deadly virus that severely affected the animal population in China.

By late 2007, an unusually high increase in very serious adverse effects associated with heparin treatments, including deaths, alerted the US health authorities and Baxter, the company that had produced the drugs concerned. In mid-January 2008, Baxter recalled several specific batches, then issued a nationwide recall a few days later. The FDA and Baxter were faced with a difficult decision to make, as a larger recall could lead to shortages of an important drug. Baxter sourced the active principle, raw heparin, from a Wisconsin company, Scientific Protein Laboratories, which manufactured half of its heparin from Midwestern pork and sourced the other half from a Chinese subcontractor, Changzhou SPL. In mid-February, the *New York Times* revealed that the FDA had never inspected this subcontractor (Bogdanich and Hooker 2008; Harris 2008).

This gave rise to a public discourse of regulatory negligence, which formed the basis of a subsequent political inquiry. The FDA organised scientific coordination to investigate issues with batches causing anaphylactic reactions; within a few weeks, this led to a consensus on the presence of an undetermined contaminant that resembled heparin. In parallel, in early March, German authorities also reported the recall of heparin batches following side effects resembling those of the US cases. These batches were linked to another raw heparin supplier, suggesting that contamination had occurred in the supply chain upstream of the Chinese subcontractor (since batches from different suppliers caused the same side effects). Recalls were issued in a dozen countries worldwide. By mid-March, the contaminant was formally identified as oversulphated chondroitin sulphate, a substance resembling heparin that is quite rare naturally, and that could not have been a result of the heparin purification process, at least not in the very high concentrations found in some samples (Guerrini et al. 2008). This array of evidence, combined with the fact that this specific contaminant could not be detected by outdated reference pharmacopeia tests, prompted the hypothesis of voluntary contamination. A month later, experiments with animal models linked the substance to the type of allergic reactions observed in patients, thus articulating the analytical hypothesis with physiological consequences (Kishimoto et al. 2008). Ultimately, within a few weeks from the first clinical reports to the recalls and then the development of analytical tests to identify contaminated batches, this major pharmaceutical scandal caused approximately 200 deaths, mostly in the USA and Germany.

The explanation for the batches' contamination upstream in the supply chain is linked to an animal pandemic. Earlier, in June 2006, signs of the spread of a porcine disease commonly called blue ear disease had appeared in China. The Porcine Reproductive and Respiratory Syndrome Virus (PRRSV), first recognised in the late 1980s in the USA, causes high mortality among piglets, and among adult pigs to a lesser extent. Vaccines have been developed, but the virus variants can evolve genetically in highly diverse ways and can thus escape immunity[2]. As with BSE, this epidemic was inextricably linked to industrialised farming. The large-scale emergence of this virus after a long evolutionary history was tied to the growth of high-density pig farms in the second half of the twentieth century, as several factors 'radically alter[ed] the ecological niche of PRRSV and facilitate[d] an explosive evolutionary radiation' (Murtaugh et al. 2010). The emergence of a new virulent variant induced the 2006 epidemic, which spread across half of China and, according to several teams of Chinese researchers, affected more than 2 million pigs over the summer of 2006 with a 20% death toll, amounting to 400,000 individuals (Tian et al. 2007). The virus kept circulating, and a new outbreak developed over the following summer. Circulation of the virus was now reported in most Chinese provinces, with journalistic investigations reporting numbers of devastated herds seemingly incompatible with the official figures quoted above (Barboza 2007; Cha 2007). At the same time, the numbers reported by the Chinese Ministry of Agriculture and by industry were wildly different, ranging from thousands to millions[3]. The epidemic had a noticeable effect on wholesale pork prices and consequently on the price of crude and refined heparin, which more than doubled between May and November 2007[4]. As oversulfated chondroitin sulphate is far cheaper than heparin, voluntary contamination was thus economically motivated: oversulfated chondroitin sulphate was used as a substitute to reduce costs in the context of rapidly rising heparin prices.

This episode unfolded at the crossroads between this emerging disease and pharmaceutical globalisation with the turn towards China during the 2000s. In the specific case of heparin, the turn towards China was dictated by the need to find new abundant pig sources following the BSE crisis. The heparin scandal was a milestone in debates on fake medicines and pharmaceutical globalisation. While the scandal of adulterated heparin occurred in the still dominantly trium-phant stage of globalisation preceding the 2008 financial crisis, for many it was a harsh wake-up call regarding the rapid mass displacement of pharmaceutical production to China and, more generally, the difficulty of monitoring increas-ingly complex global supply chains that rely on outsourcing and subcontractors. In the United States, critical perspectives on the uncontrolled consequences of pharmaceutical globalisation and the difficulties associated with its regulation started to crystallise within the discourses of elected officials as well as FDA scientists and administrators. Testifying at a hearing in the House of Representatives, Janet Woodcock, director of the Center for Drug Evaluation and Research at the FDA, stated:

> The sites of production of pharmaceuticals have changed. [...]. Over the past 15 years, the majority of active pharmaceutical ingredient manufacture and actually increasing amounts of finished drug product manufacture has moved off our shores, been outsourced. For example, generic drug applications processed in 2007 at the FDA referenced over 1,000 foreign sites; 450 of those were in India, 497 of those were in China for API manufacture of those generic drugs. And only 151 of them were in the United States. The rest were in other countries around the world. The FDA of the last century is not configured to regulate this century's globalised pharmaceutical industry (Committee on Energy and Commerce, 2008: 45).

Alarmed commentary proliferated in the public sphere. To quote one of many examples:

> Like the canary that stops singing in the coal mine, the heparin safety crisis is an early and urgent warning of the vulnerability and impending failure of our drug safety system. We must respond to this warning with an extreme sense of urgency if we are to prevent more catastrophic failures of this system as a result of either mistaken or intentional drug adulteration. (Leiden 2008: 626)

In most discourses, however, the solution identified was stricter regulation to prevent the dire consequences that could ensue from the ill-intentioned exploitation of the seemingly inevitable globalisation trend. While the heparin scandal shed stark light on the growing issue of counterfeit drugs, it also provided a ready-made solution to rescue the globalisation process – adaptation through regulation – rather than questioning the outsourcing trend and increasingly complex supply chains (Quet et al. 2018).

While the FDA was harshly criticised for failing to regulate, it was also praised for its rapid response, which prevented more deaths. The FDA was able to leverage the shocked reactions to the scandal to gain more regulatory power, particularly with the FDA Globalisation Act of 2009. In parallel, it coordinated the reviewing and strengthening of quality control testing for heparin through the development of new tests (Szajek et al. 2016).

In this episode, a set of unknown actors, upstream in a globalised supply chain, locally adapted to a large-scale pig epidemic and used technical biochemical knowledge on heparin production and regulatory testing to make a profit by introducing a deadly fraud in the commercial circuits of heparin. The dire consequences, for patients, had worldwide ripple effects on the regulation of drugs. While one can (and should) point out the dreadful impact that adulterated drugs can have on patients downstream, these consequences were also the result of an uncontrolled globalisation process in which regulation always tends to be introduced only after the realisation that something has gone badly wrong. In this particular case, the pharmaceutical companies that produced finished heparin drugs had until then been free to source animal matter from wherever it could be found:

Interviews with dozens of heparin producers and traders in several Chinese provinces, as well as a visit to a village near here dominated by tiny family workshops that process crude heparin from pig intestines, show the difficulties confronting investigators as they seek to trace the supply chain. [...] The Chinese heparin market has become increasingly unsettled over the last year, as pig disease has swept through the country, depleting stocks, leading some farmers to sell sick pigs into the market, and forcing heparin producers to scramble for new sources of raw material. Traders and industry experts say even big companies have been turning more often to the small village workshops, which are unregulated and often unsanitary. [...] Some experts say as much as 70 percent of China's crude heparin for domestic use and export comes from small factories in poor villages. One of the biggest areas for these workshops is here in coastal Jiangsu Province, north of Shanghai, where entire villages have become heparin production centers. In a village called Xinwangzhuang, nearly every house along a narrow street doubles as a tiny heparin operation, where teams of four to eight women wearing aprons and white boots wash, splice, separate, and process pig intestines into sausage casings and crude heparin. (Barboza and Bogdanich 2008)

Following the scandal, the FDA launched another move to reorganise the heparin supply chain and rewire the circuits of globalisation to disentangle them from China, through the coordination of expertise from both the Global North and the Global South. A discussion to reintroduce heparin sourced from bovine materials took place in June 2015, at an FDA Science Board Public Meeting (Al-Hakim 2021). Speakers voiced their concerns regarding the stability of a supply that relied mostly on one major source, Chinese pigs, pointing either to the case of porcine disease or to geopolitical instability. Reintroducing bovine sources could provide a way to broaden the geographic distribution of the animal resources needed and reduce the risk of shortages. The first issue surrounding the reintroduction of bovine heparin was how to compare its activity profiles with heparin of porcine origin: for example, as bovine heparin is less active, doses have to be adjusted. Moreover, side effects, notably allergies, needed to be taken into account. Broadly speaking, the most recent research pertained to porcine heparin. The second issue was the possibility of contamination by prions. Proponents of bovine reintroduction, such as the FDA, advocated the reevaluation of biosecurity risks: the significant amount of knowledge on prion inactivation gained since the BSE crisis could be incorporated into heparin production processes to reduce contamination risks. The use of cattle coming from countries free of the disease could further reduce risks. The National Institute for Biological Standards and Control (UK) and United States Pharmacopeia organised workshops on the characterisation of heparin-based products in 2015 and 2017, notably convening with experts from countries that had maintained the use of bovine heparin, such as Brazil, Argentina, and India. Global South expertise is thus being used in the current reorganisation and diversification of global heparin circuits.

EPISODE 3: SHORTAGE FEARS AND SYNTHETIC HOPES

Within a few years, the risk of heparin shortages soon materialised, due to yet another animal epidemic that dealt a massive blow to Chinese pig herds and stretched the heparin supply thin. This last episode relates to African swine fever (ASF). As SARS-CoV-2 swept across the globe in early 2020, various Chinese animals became the centre of attention worldwide. But starting in 2018 and 2019, another developing pandemic, African swine fever, took hold in the country and swept through pig herds. In 2010, the PRRS virus had been termed 'the most significant swine disease worldwide in spite of intensive immunological interventions' (Murtaugh et al. 2010: 18) because of its high economic impact. Less than ten years later, these recurring outbreaks were dwarfed by the sheer scale of the new pandemic, which kept expanding worldwide. ASF spreads in factory-farm pigs, but also in wild boars: the countermeasures that are applied in Europe to try to stop the spread are thus extending veterinary logic to the ways in which human-wildlife interactions are managed (Broz et al. 2021). Estimates vary on the number of pigs killed either by the virus or in preventive culling, but the figure is probably in the several hundreds of millions[5].

ASF sparked fears of a potentially massive heparin shortage, causing many local shortages and supply chain tensions. This revealed the fragility of readily available resources (pig numbers are one thing, but the animal materials also need to be collected and integrated into complex circuits of exploitation for heparin production), and strongly supported the call to reorganise the supply chain and make it more flexible (Fareed et al. 2019; McCarthy et al. 2020; Rosovsky et al. 2020). Several countries, including France, the USA, and Brazil, are currently building new factories to manufacture raw heparin out of local sources in order to stabilise the supply. In the wake of this episode, which was further compounded by the massive global supply chain disruptions caused by COVID-19, pharmaceutical globalisation is now being directly called into question.

China continues to supply worldwide heparin, and some of its large agro-industrial companies are using the aftermath of the devastation caused by ASF to advocate gigantic new pig farms as the new global frontier of technologised exploitation of animal life, creating a breeding site for future pandemics[6].

Meanwhile, the supply chain of Chinese raw heparin for companies of the Global North still exists, but decades of expertise on heparin is now leading to the formation of large companies integrating the entire heparin supply chain, from animals to finished drug products. These companies produce biopharmaceuticals for the Global South and, more recently, for markets in the Global North. Additionally, they are exploring the potential development of new pharmaceuticals derived from this century-old technique[7].

Since the 2008 scandal, and with greater urgency lately following the spread of ASF in China, there have been renewed incentives for the development of synthetic heparin.

> To get around the problem that sourcing heparin from animals poses, one solution could be synthetic heparin or a synthetic drug that performs like heparin. Several groups of researchers around the world are trying to do just that, despite the long odds against succeeding. 'When we first started, we tried to make heparin itself', recalls Jian Liu, a professor at the Eshelman School of Pharmacy at the University of North Carolina, Chapel Hill, who heads a group that has been trying to synthesise heparin for more than a decade. 'But how can you reproduce a molecule that cannot be precisely characterised and that also contains 40 sugars?' (Tremblay 2016)

The extremely complicated chemistry of synthesising massive heparin molecules might be solved by focusing on the activity of smaller molecules. Some of them, such as fondaparinux, already exist, but they are not yet suitable for all the clinical applications of other heparin molecules. It is also hoped that biotechnological engineering could finally help design heparin that is separated from its animal origins (Baytas and Linhardt 2020). The biotechnological fix could very well come about soon and add to the list of biomedical innovations that have reduced human dependence on animals for therapeutic needs, such as insulin. But it may also never be fully successful, and the relationship between animals and human health could remain, raising a set of issues briefly outlined in the conclusion.

CONCLUSION: THE ENVIRONMENTAL SUSTAINABILITY OF THERAPEUTIC FUTURES

The troubled historical and geographical life of heparin, from animal guts to human medicine, helps reveal the unseen connections between biomedical globalisation and the combined diseases of humans and non-human animals alike. By analysing episodes when silent actors came to the forefront of the global stage, this chapter has shown that exploited animals can also be direct players in globalisation – especially when they get sick (Nading 2013). Furthermore, this case study contradicts the widespread view of Southern countries and people as the overpowered figures of globalisation, and that of an ever-growing asymmetry of power between North and South, with the former reinforcing its power over the latter. As regards the globalisation of pharmaceuticals, it questions the image of the linear relationship of Northern-based Big Pharma looking at the South as mere emerging markets to conquer (or mere sites for raw material extraction). It also shows that the effects of globalisation on the construction of global supply chains in the food and pharmaceutical industries, as well as the effects of emerging diseases provoked by human action, may have the unexpected consequence of reversing global power configurations, with actors supposedly from below playing an active and privileged role that may define pharmaceutical and economic policies in the North. In that perspective, heparin is helping us to nuance our understanding of globalisation processes.

But there is even more to this story. Heparin challenges our definition of waste. This chapter started with a disease that originated from the circulation of waste, in the form of processed animal matter that had been fed to cattle in Britain, thus allowing a crossing of species barriers and the emergence of BSE. The uses of animal waste took on new proportions with the emergence of the slaughterhouse industry, with a drive to make the byproducts of food production profitable in any way possible. The existence of heparin as a drug is a consequence of the ready availability of animal waste in large quantities, as a result of the rise of industrial farming in the twentieth century. However, approaching the production of this drug as a useful outlet for matter (intestinal mucus) that otherwise serves no purpose makes it possible to ignore the environmental externalities associated with industrial farming. Heparin appears as an almost free resource, a derivative of food production processes that are already in place, and as such having no environmental footprint, or even a positive one (notwithstanding the large amounts of chemical solvents used as part of the extraction process) since it puts to use matter that would otherwise be lost.

The fears of a heparin shortage are turning this perspective on its head: What if the waste were to become the primary resource? What if, due to an epidemic or geopolitical events, pig intestines were to stop being available? The rapid succession of epidemics that has threatened heparin production may well continue to develop, and if technological solutions fail, we may still be dependent on cows and pigs for its supply for some time to come. What is the minimum number of animals needed in order to continue to produce this essential drug? Is the sustained supply of heparin compatible with a reduction of intensive farming, either forcibly because of epidemics or in a controlled way to ensure environmental sustainability? These questions become more pressing if we further consider the dimension of ethical access to drugs. Today, shortages are also linked to increased consumption due to a growing and aging population, especially in countries of the Global North, where most heparin is used.

These questions also rely on the shape that has been given to global economic exchanges during the last 40 years. Whereas Chinese pig farmers appeared in the 1990s as obvious raw material providers, the procurement crises described above obliged the richest countries to revise this assumption. Shifting conceptions of dependence and international competition have turned questions of heparin supply into a geopolitical issue. In that evolving context, the globalisation of pharmaceutical raw material supply chains has been submitted to increasing criticism. Heparin procurement offers a striking example of the reconfiguration of power relations that has been taking place during the recent globalisation phase – which might complicate the idea of globalisation both 'from above' and 'from below', offering an illustration of how economies of scale and global outsourcing have also collided with biomedical constraints and geopolitical power considerations. This could inspire further social science research that considers how seemingly dominated actors, from Global South countries and their populations to non-humans, can trigger economic and political changes at the global scale.

ENDNOTES

1 Anipharm (Animal Products and Biomedical Globalisation from the South: The pharmaceutical uses of animal life in the Indian Ocean) is an ANR-funded project coordinated by Mathieu Quet from 2020 to 2024.

2 Wageningen University & Research. (2019, November 4). Blue Ear Disease (PRRS). https://www.wur.nl/en/research-results/research-institutes/bioveterinary-research/animal-diseases/virology/prrs-or-blue-ear-disease.htm.

3 The Pig Site. (2007, August 21). The Story Behind China's Rising Pork Prices. https://www.thepigsite.com/articles/the-story-behind-chinas-rising-pork-prices.

4 PEW Health Group. (2011). After Heparin: Protecting Consumers from the Risks of Substandard and Counterfeit Drugs. https://www.pewtrusts.org/en/research-and-analysis/reports/2011/07/12/after-heparin-protecting-consumers-from-the-risks-of-substandard-and-counterfeit-drugs

5 Standaert, M. "Unstoppable": African Swine Fever Deaths to Eclipse Record 2019 Toll. *The Guardian*, Environment, (2020, May 27), https://www.theguardian.com/environment/2020/may/27/unstoppable-african-swine-fever-deaths-to-eclipse-record-2019-toll.

6 Standaert, M., & De Augustinis, F., 'A 12-Storey Pig Farm: Has China Found the Way to Tackle Animal Disease?' *The Guardian*, Environment., (2020, September 18). https://www.theguardian.com/environment/2020/sep/18/a-12-storey-pig-farm-has-china-found-a-way-to-stop-future-pandemics-.

7 Shenzhen Hepalink Pharmaceutical Group Co., Ltd. Accessed 23 March 2023. https://www.hepalink.com/en/index.aspx.

REFERENCES

Agence Nationale de Sécurité du Médicament, *État des lieux sur les heparines* (2014).

Al-Hakim, A., 'General Considerations for Diversifying Heparin Drug Products by Improving the Current Heparin Manufacturing Process and Reintroducing Bovine Sourced Heparin to the US Market', *Clinical and Applied Thrombosis/Hemostasis*, 27 (2021): 107602962110522 https://doi.org/10.1177/10760296211052293.

Anderson, J. L., *Capitalist Pigs: Pigs, Pork, and Power in America* (West Virginia University Press, 2019).

Anderson, W., *The Collectors of Lost Souls: Turning Kuru Scientists into Whitemen* (Johns Hopkins University Press, 2008).

Barboza, D., 'Virus Spreading Alarm and Pig Disease in China', *The New York Times*, 16 August 2007 https://www.nytimes.com/2007/08/16/business/worldbusiness/16pigs.html.

Barboza, D., and W. Bogdanich, 'Twists in Chain of Supplies for Blood Drug', *The New York Times*, 28 February 2008 https://www.nytimes.com/2008/02/28/world/asia/28drug.html.

Barrowcliffe, T. W., 'History of Heparin', in R. Lever, B. Mulloy, and C. P. Page, eds., *Heparin – A Century of Progress: Handbook of Experimental Pharmacology* (Springer, 2012), pp. 3–22 https://doi.org/10.1007/978-3-642-23056-1.

Baytas, S. N., and R. J. Linhardt, 'Advances in the Preparation and Synthesis of Heparin and Related Products', *Drug Discovery Today*, 25.12 (2020): 2095–2109 https://doi.org/10.1016/j.drudis.2020.09.011.

Blanchette, A., *Porkopolis: American Animality, Standardised Life, and the Factory Farm* (Duke University Press, 2020).

Blattner, C., *Protecting Animals Within and Across Borders* (Oxford University Press, 2019).

Bogdanich, W., and J. Hooker, 'China Didn't Check Drug Supplier, Files Show', *The New York Times*, 16 February 2008 https://www.nytimes.com/2008/02/16/us/16baxter.html.

Browaeys, D. B., 'Anticoagulants: La Filière Chinoise', *60 Millions de Consommateurs* (November 2005).

Brown, H., and A. M. Nading, 'Introduction: Human Animal Health in Medical Anthropology', *Medical Anthropology Quarterly*, 33.1 (2019): 5–23 https://doi.org/10.1111/maq.12488.

Broz, L., A. Garcia Arregui, and K. O'Mahony, 'Wild Boar Events and the Veterinarization of Multispecies Coexistence', *Frontiers in Conservation Science*, 2 (December 2021): 711299 https://doi.org/10.3389/fcosc.2021.711299.

Cha, A. E., 'Pig Disease in China Worries the World', *The Washington Post*, 16 September 2007 http://www.washingtonpost.com/wp-dyn/content/article/2007/09/15/AR2007091501647.html.

Chee, L. Y., *Mao's Bestiary: Medicinal Animals and Modern China* (Duke University Press, 2021).

Chess, E. K., S. Bairstow, S. Donovan, and others, 'Case Study: Contamination of Heparin with Oversulfated Chondroitin Sulfate', in R. Lever, B. Mulloy, and C. P. Page, eds., *Heparin – A Century of Progress: Handbook of Experimental Pharmacology* (Springer, 2012), pp. 99–125 https://doi.org/10.1007/978-3-642-23056-1.

Clarke, A. E., 'Research Materials and Reproductive Science in the United States, 1910–1940 with Epilogue: Research Materials (Re) Visited', in S. L. Star, ed., *Ecologies of Knowledge: New Directions in Sociology of Science and Technology* (State University of New York Press, 1995), pp. 183–225.

Collard, R.-C., *Animal Traffic: Lively Capital in the Global Exotic Pet Trade* (Duke University Press, 2020).

Committee on Energy and Commerce. *The Heparin Disaster: Chinese Counterfeits and American Failures*. House of Representatives. (2008) Hearing Serial No. 110-109. https://www.govinfo.gov/content/pkg/CHRG-110hhrg53183/html/CHRG-110hhrg53183.htm.

Coyne, E., 'Heparin – Past, Present and Future', in R. L. Lundblad, W. V. Brown, K. G. Mann, and H. R. Roberts, eds., *Chemistry and Biology of Heparin* (Elsevier, 1981), pp. 9–18.

Craddock, S., and S. Hinchliffe, 'One World, One Health? Social Science Engagements with the One Health Agenda', *Social Science & Medicine*, 129 (2015): 1–4 https://doi.org/10.1016/j.socscimed.2014.11.016.

Fareed, J., W. Jeske, and E. Ramacciotti, 'Porcine Mucosal Heparin Shortage Crisis! What Are the Options?', *Clinical and Applied Thrombosis/Hemostasis*, 25 (2019): 1076029619878786 https://doi.org/10.1177/1076029619878786.

Frutos, R., and others, 'Emerging Infectious Diseases in Ungulates', *Frontiers in Veterinary Science*, 8 (2021): 620691 https://doi.org/10.3389/fvets.2021.620691.

Gao, H., and others, 'Diversification of Heparin Products: Past, Present, and Future', *Thrombosis and Haemostasis*, 121.2 (2021): 125–36 https://doi.org/10.1055/s-0040-1716862.

García-Sancho, M., *Animal Breeding in the Age of Biotechnology: The Investigative Pathway Behind Dolly the Sheep* (Springer Nature, 2021).

Garritano, J., *Vital Strife: Humanity and Its Discontents* (Cornell University Press, 2020).

Greenwood, B., *The Microbe as a Metaphor: Scientific Language and Medical History* (University of Chicago Press, 2019).

Hamlin, C., *Cholera: The Biography* (Oxford University Press, 2009).

Hess, D. J., 'Medical Modernisation, Scientific Research Fields, and the Epistemic Politics of Health Social Movements', *Sociology of Health and Illness*, 26.6 (2004): 695–709 https://doi.org/10.1111/j.1467-9566.2004.00414.x.

Hinchliffe, S., and N. Bingham, 'Securing Life: The Emerging Practices of Biosecurity', *Environment and Planning A*, 40.7 (2008): 1534–51 https://doi.org/10.1068/a4054.

Holmes, M. C., 'Heparin and Its Production', in R. Lever, B. Mulloy, and C. P. Page, eds., *Heparin – A Century of Progress: Handbook of Experimental Pharmacology* (Springer, 2012), pp. 1–21 https://doi.org/10.1007/978-3-642-23056-1_1.

Hooper, M. J., and others, 'Animal Models of Emerging Human Disease: An Overview of the Field', *Comparative Medicine*, 55.3 (2005): 221–31.

Houck, O. A., *Taking Back Eden: Eight Environmental Cases That Changed the World* (Island Press, 2009).

Kirksey, S. E., *Emergent Ecologies* (Duke University Press, 2015).

Knapp, C. R., and others, 'Wildlife Markets and Zoonotic Disease Risk in Laos', *Frontiers in Veterinary Science*, 7 (2020): 558172 https://doi.org/10.3389/fvets.2020.558172.

Krishna, V. N., and others, 'How We Got to Now: Historical Evolution of Heparin', *Seminars in Thrombosis and Hemostasis*, 45.3 (2019): 217–21 https://doi.org/10.1055/s-0039-1684021.

LaCapra, D., 'Globalization and Critical Animal Studies', pp. 136–154 in Amanda Minervini, Amelie Björck, Omri Grinberg, and Amrita Ghosh (Eds) *ReFiguring Global Challenges. Literary and Cinematic Explorations of War, Inequality, and Migration* (Brill, 2023)

Lakoff, A. 'The generic biothreat, or, how we became unprepared' *Cultural Anthropology*, 23(3) (2008): 399-428.

Latour, B., *We Have Never Been Modern* (Harvard University Press, 1993).

Lawrence, T., *The Heparin Story: A Golden Anniversary Celebration* (Springer, 1988).

Lee, P. Y. (Ed.). *Meat, Modernity, and the Rise of the Slaughterhouse* (University of New Hampshire Press, 2008)

Leiden, J. M., 'Canaries, Coal Mines and the Drug Supply', *Nature Biotechnology*, 26(6) (2008): 624–626. https://doi.org/10.1038/nbt0608-624.

Lever, R., and B. Mulloy, 'Structure and Biology of Heparin', in R. Lever, B. Mulloy, and C. P. Page, eds., *Heparin – A Century of Progress: Handbook of Experimental Pharmacology* (Springer, 2012), pp. 45–71 https://doi.org/10.1007/978-3-642-23056-1_2.

Liang, L. L., and others, 'The Role of Bovine Heparin in the Diversification of Anticoagulant Therapies', *Thrombosis and Haemostasis*, 121.4 (2021): 450–60 https://doi.org/10.1055/s-0040-1717364.

Linhardt, R. J., 'Heparin: Structure and Activity', *Journal of Medicinal Chemistry*, 46.13 (2003): 2551–64 https://doi.org/10.1021/jm020566b.

Linhardt, R. J., and C. L. Gunay, 'Production and Chemical Processing of Heparin', in R. L. Lundblad, W. V. Brown, K. G. Mann, and H. R. Roberts, eds., *Chemistry and Biology of Heparin* (Elsevier, 1981), pp. 57–72.

Linhardt, R. J., and others, 'Heparin: An Overview', *Acta Poloniae Pharmaceutica – Drug Research*, 60.3 (2003): 243–47.

Luo, J., and others, 'The Significance of Heparin in Medical Practice', *Thrombosis Research*, 158 (2017): 126–32 https://doi.org/10.1016/j.thromres.2017.07.017.

Mackie, A., and others, 'Analysis of Heparin Contamination Events', *Emerging Infectious Diseases*, 13.12 (2007): 1910–13 https://doi.org/10.3201/eid1312.070303.

Marder, E. M., and others, *Heparin: Structure, Function, and Therapeutic Potential* (Oxford University Press, 2014).

Martin, C., 'The Ethics of Animal Research in the Context of Heparin Production', *Journal of Bioethical Inquiry*, 9.3 (2012), 317–28 https://doi.org/10.1007/s11673-012-9390-6.

McCarthy, C. P., Vaduganathan, M., Solomon, E., Sakhuja, R., Piazza, G., Bhatt, D. L., Connors, J. M., & Patel, N. K. 'Running Thin: Implications of a Heparin Shortage', *The Lancet, 395*(10223) (2020): 534–536. https://doi.org/10.1016/S0140-6736(19)33135-6.

Meyer, R., *Unsettling Science: The Politics of Heparin in the Global Age* (Routledge, 2017).

Mills, E., and others, 'Risk Assessment of Heparin and Heparin-Based Products', *Thrombosis Research*, 126.5 (2010): 439–47 https://doi.org/10.1016/j.thromres.2010.06.004.

Murtaugh, M. P., Stadejek, T., Abrahante, J. E., Lam, T. T. Y., & Leung, F. C.-C., 'The Ever-Expanding Diversity of Porcine Reproductive and Respiratory Syndrome Virus', *Virus Research, 154*(1–2) (2010): 18–30. https://doi.org/10.1016/j.virusres.2010.08.015.

Nading, A., 'Humans, Animals, and Health: From Ecology to Entanglement' *Environment and Society: Advances in Research* 40(1) (2013): 60-78

Oudshoorn, N., *Beyond the Natural Body: An Archeology of Sex Hormones* (Routledge, 1994)

Page, C. P., and others, 'Heparin and Its Role in the Blood Coagulation Cascade', *Blood Coagulation & Fibrinolysis*, 29.7 (2018): 538–48 https://doi.org/10.1097/MBC.0000000000000744.

Park, M., Singer. P., 'The Globalization of Animal Welfare: More Food Does Not Require More Suffering', Foreign Affairs, 91, (2012): 122.

Parker, H., *Zoonotic Diseases: Challenges in the Control of Emerging Pathogens* (Springer, 2018).

Phillips, A., and others, 'Heparin and Other Anticoagulants in Cardiovascular Therapy', *Journal of Clinical Pharmacology*, 56.5 (2016): 543–52 https://doi.org/10.1002/jcph.720.

Reiss, H., *A History of Heparin and Its Discovery* (Cambridge University Press, 2021).

Rosovsky, R. P., Barra, M. E., Roberts, R. J., Parmar, A., Andonian, J., Suh, L., Algeri, S., & Biddinger, P. D. 'When Pigs Fly: A Multidisciplinary Approach to Navigating a Critical Heparin Shortage' *The Oncologist, 25*(4) (2020): 334–347. https://doi.org/10.1634/theoncologist.2019-0910.

Sands, A., and others, 'Investigating the Safety and Efficacy of Heparin in Clinical Practice', *Journal of Clinical Hematology*, 18.3 (2019): 185–91 https://doi.org/10.1016/j.jclinhem.2019.01.003.

Schwerin, A. von, Stoff, H., & Wahrig, B. (Eds.) *Biologics, A History of Agents Made From Living Organisms in the Twentieth Century* (Pickering & Chatto, 2013).

Smith, M., and others, *The Age of Heparin: Medical History and Development* (Springer, 2016).

Specht, J., *Red Meat Republic: A Hoof-to-Table History of How Beef Changed America* (Princeton University Press, 2020)

Sykes K., 'Globalization and the Animal Turn: How International Trade Law Contributes to Global Norms of Animal Protection', Transnational Environmental Law. 5(1) (2016):55-79.

Szajek, A. Y., Chess, E., Johansen, K., Gratzl, G., Gray, E., Keire, D., Linhardt, R. J., et al. The US Regulatory and Pharmacopeia Response to the Global Heparin Contamination Crisis. *Nature Biotechnology, 34*(6) (2016): 625–630. https://doi.org/10.1038/nbt.3606.

Taylor, R., *Biotechnology and the Rise of Heparin: Medical Impact and Cultural Shifts* (Sage, 2008).

Thompson, W., *Modern Medical Practices and Heparin* (Beacon Press, 2020).

Tian, K., Yu, X., Zhao, T., Feng, Y., Cao, Z., Wang, C., Hu, Y., et al. 'Emergence of Fatal PRRSV Variants: Unparalleled Outbreaks of Atypical PRRS in China and Molecular Dissection of the Unique Hallmark.', *PLoS ONE, 2*(6), e526. (2007). https://doi.org/10.1371/journal.pone.0000526.

Tremblay, J.-F., 'Making Heparin Safe' *C&EN Global Enterprise, 94*(40), 2016: 30–34. https://doi.org/10.1021/cen-09440-cover.

Williams, D., and others, 'In Vitro Activity of Heparin Derivatives in the Prevention of Venous Thromboembolism', *Clinical and Applied Thrombosis/Hemostasis*, 19.1 (2013): 49–55 https://doi.org/10.1177/1076029612462510.

Wilson, S., and others, 'Heparin: From its Discovery to Modern Medical Applications', *Journal of Anticoagulant Therapy*, 21.4 (2017): 425–35 https://doi.org/10.1016/j.anticoag.2017.02.001.

Wright, M., and others, 'Heparin and Heparin-Like Substances: Biological Implications for Medical Practice', *Scientific Reports*, 7.1 (2018): 3911 https://doi.org/10.1038/s41598-017-04567-9.

Zhang, M., Bjerke, L., 'Antibiotics "dumped": Negotiating Pharmaceutical Identities, Properties, and Interests in China–India Trade Disputes', *Medical Anthropology Quarterly*, 37/2 (2023): 148-163

Zhu, W., and others, 'Therapeutic Uses of Heparin in Surgery', *Journal of Cardiothoracic and Vascular Anesthesia*, 29.5 (2015): 1187–95 https://doi.org/10.1053/j.jvca.2015.01.023.

ADJUSTING THE GLOBAL

7

PATCHING DEVELOPMENT: INFORMATION TECHNOLOGY ADJUSTMENTS IN THE MAURITIAN LOGISTICAL SECTOR

Marine Al Dahdah and Mathieu Quet

IN FEBRUARY 2021, THE DECATHLON GROUP UNVEILED AN ENORMOUS 26,000 square meter warehouse on the island of Mauritius to serve as a logistical hub between Asia, where most of the company's products are manufactured, and Sub-Saharan Africa, the Indian Ocean, and the Middle East – regions whose emerging middle classes represent a growing share of the brand's consumers. During an official visit to the site, the Mauritian Minister of Finance, Renganaden Padayachy, reminded the audience that the choice of the island (over competing locations such as Dubai and Durban) reinforced the importance of Mauritius as an ideal hub for investors and for the continued export of goods to other countries in the region. The president of Mauritius Free Port Development (MFD), which manages the warehouse, emphasised that the event was significant not only because of the volume of containers and goods flowing in and out, but also for Mauritians themselves, because of the transfer of expertise and knowledge the warehouse's operation would entail.

As these speeches made explicit, such an event points to one particular development strategy adopted by states and firms in the context of industrial

FIG. 7.1 Along the Mauritius Free Port Zone, January 2020 (Mathieu Quet)

capitalism, which consists in attracting and redistributing flows of goods they do not manufacture, in order to benefit from the surplus value generated during transport and storage operations. This logistical engagement with capitalism aims at capturing value through the management of flows. Although it is not new, this strategy has assumed growing importance during the last decades, owing to the ever-expanding volume of goods traded globally; in return, it has increasingly contributed to shape the material and social manifestations of globalisation, through the multiplication of warehouses, megaports, megaships, and freezones that populate developing landscapes. Mauritius provides an interesting illustration of this phenomenon when one takes into account the various policies shaped by successive Mauritian governments over the past several decades to attract investment and goods flows – what some authors refer to as 'concentration' or 'hubbing' (Schnepel 2018). Indeed, since its independence – and especially since the 1990s – Mauritius has developed various incentive structures for foreign companies, beginning with the creation of a free port in 1992 and continuing with the development of offshoring and telecommunications services (maintenance, call centres). This policy has been part of a forward-looking conceptualisation on the part of the government and private sector actors, and it benefitted from the support of the country's Economic Development Board. It aimed at overcoming Mauritius' isolated status and turning its geographical, language, and cultural characteristics into decisive assets over other destinations.

One essential way to achieve such a strategy and to successfully attract flows is to offer to make them 'seamless', as promoted on the MFD website. In Deborah Cowen's analysis of contemporary logistics, she observes that 'there is a growing common sense that the competitiveness of firms, nations, and supranational regions is contingent on their capacity to mobilise "seamless" supply chains, to circulate stuff in a timely and reliable way' (2014: 58). According to this logistical understanding, seamlessness signifies the quality of flows moving efficiently and smoothly, without any disruption or slowdown – an objective that can never fully be achieved but fuels the imagination and normative approaches of what would be an ideal supply chain.

This leads us to the following research question: How can one provide, create, or manufacture seamlessness? What are the material, organisational, epistemic ways of fabricating global, yet smooth, flows of goods? Sandro Mezzadra and Brett Neilson (2019) describe as 'operations' the multiple ways or tools through which flows are managed, manipulated, and integrated: Enterprise resource planning software, cranes, containers, ports, and free trade agreements then appear as multiple and heterogeneous operations aiming to facilitate and accelerate the flows of goods and their transformation into flows of capital. These authors submit the hypothesis that the 'concatenation' of operations is one of the central challenges in technological capitalism today: creating homogeneity between operations of capital means making them interoperable and entails the constant production of intercommunication. Mezzadra and Neilson's intervention opens an avenue for research as it invites analysis of how operations and their concatenation manifest 'on the ground' and contribute to the globalisation of trade.

In this perspective, the Mauritian development model (Chazan-Gillig and Widmer 2001; Grégoire 2016) sheds interesting light on this phenomenon as it illustrates an essential trait of capitalist revitalisation strategies to attract flows and capture value: the increasing integration of logistics and digital technology.

This is what we observed over the course of two research trips (in January 2020 and January 2022) during which we conducted two series of interviews: first (S1, 2020: 17 interviews) with various actors from the Mauritian digital sector, and second (S2, 2022: 25 interviews) with a company specialising in information technology services and solutions for the logistics sector. Our study allowed us to document 'operational' capitalism in the convergence between the digital and logistical sectors. However, it also led us to a further analysis, beyond Mezzadra and Neilson's initial hypothesis: while the imperative to produce 'seamless' flows is shared by all actors in international logistics (Cowen 2014), Mauritian actors have had to implement specific strategies due to their resource levels, geographic position, and target markets. We will focus in particular on the concept of 'patching,' which refers to the fact that achieving fluid circulation requires a kind of technological bricolage, and the continuous adaptation of information systems that must be fitted and adjusted in order to respond to local constraints. The manufacture of seamless flows, observed through the lens of the convergence of Mauritian logistics and information technology (IT), requires a particular kind of customisation work, the innovative nature of which must be better assessed in order to more fully describe the integration of developing countries into global capitalism. Furthermore, patching operations shed a particular light upon adjustment and adaptation practices as a recurring feature of the experience of technological globalisation in the Global South. Patching is both a process required by 'seamlessification' and an answer to the great heterogeneity of IT solutions involved by the globalisation of markets. On one hand, the Mauritian IT sector is highly dependent on IT systems often imposed by the most powerful actors of the logistical industry; on the other hand, the lower level of resources of clients operating in Mauritius' trade area often requires finding cost-effective solutions in order to adjust dominant IT systems to cheaper, tailored situations. In that sense, patching simultaneously appears as a prominent task in the fabrication of global logistics and as a specific answer to the coexistence of heterogeneous IT systems in terms of cost and standardisation level.

In order to make our argument, we first present the digital turn in the logistics sector as it was recounted to us by the actors we interviewed. Next, we detail the main IT activities required to create a seamless flow. And finally, we analyse the importance of the practice of patching and attempt to show its specificity within a development context.

A DIGITAL TURN IN LOGISTICS?

The main actors in our story are a logistics company and an information services company, both based in Mauritius. The history of these companies and their collaboration serves to illustrate the importance of the digital turn in the logistics industry.

Mauritius Free Port Development (MFD) is the chief private operator in charge of the development of Mauritius's free port. The project for a free port on the island was part of the economic development strategy put into place by the nation following its independence in 1968. During the years leading up to independence, the government had prepared a free zone policy by carrying out various studies of free zones in Southeast Asia and the Caribbean in order to draw inspiration from various models (Rogerson 1993). In 1970, determined to implement a free zone policy, the government of the newly independent nation passed the Export Processing Zone Act. The zone's main objective was to attract foreign investors in order to develop the island's industrial sector. It must be emphasised that until the late 1960s, Mauritius's primary export commodity was sugar cane, which accounted for 90% of all exports (Ramtohul and Erikson 2018). By reducing customs duties and labour oversight, the government sought to develop industrialisation on the island, resulting in the growth of the textile manufacturing sector over the course of the 1970s and 1980s (Neveling 2014). The free port zone created in 1992 was primarily intended to complement this original policy by facilitating exports, simplifying customs controls, and improving the island's transport and export infrastructure.

The legislation surrounding the free port offered a set of liberal incentive measures targeting businesses seeking a hub for storage, assemblage, and redistribution in the middle of the Indian Ocean. Mauritius's free port promoted its many advantages: rapid customs procedures, duty-free and VAT-free goods and services, tax-exempt status for companies, the possibility for a company's stock to be 100% held by foreigners, free repatriation of profits, no exchange controls, a 50% reduction in port handling fees, the possibility for 50% of sales revenue to come from the local market, and access to the island's offshore banking services. Along with this legal and financial infrastructure, Mauritius boasts a political and economic stability that is rare in Africa, a bilingualism (French and English) that is useful for international communications, a highly educated and trained population, and good digital connectivity, all of which are elements that further enhance the island's strategic position as an international logistical hub.

The MFD was created in 1995 within the context of the implementation of the free port zone, in the wake of a series of other institutions created to oversee the development of the port zone and the island's logistical capacities (first MMA and then MPA and MCargoHandling). MFD's role is to manage and develop the free port territory: 'MFD has a 60-year contract with the Mauritian government to develop the 25 hectares of land that make up the free port zone. Initially, MFD concentrated heavily on textiles in the early 2000s, but there were some rough years in the textile industry, so it had to reinvent itself and find other activities' (MFD IT Officer, January 2022). From seafood to medications, by way of textiles and sugar, MFD significantly diversified its activities beyond the zone of the free port, with warehouses in inland regions and at the international airport, in order to meet the growing needs of its clients. Thus, it became a company specialising in logistics and distribution, targeting the Mauritian and Indian Ocean markets, as explained by one executive: 'Our job is to receive, store, and deliver goods, and whether it's cars, pharmaceutical products, or mass-market goods, our job stays the same—it's easy to understand, but difficult to execute' (MFD IT Officer, January 2022). From a more analytic point of view, we may note that MFD's goal is first and foremost to capture goods flows by offering its services to the manufacturing firms that produce these flows and the consumption sites that consume them. This goal is apparent if we look at how MFD won the Decathlon contract, as explained to us by an MFD executive:

> One of the major projects we've had is Decathlon. To close a booster in
> Asia and open it in Mauritius. We built a special Decathlon booster where
> we handle all the Asia flows for Africa, the Middle East, and even some of

the European flows. We are able to support our clients even beyond our walls. It took some time to get it going. Decathlon corresponds precisely to the reason why the free port was created—to help these multinationals get closer to their markets; 40% of the Decathlon booster in Mauritius is East Africa and Réunion. When you look at the development strategy, it's to be as close as possible to their markets. Mauritius is a fairly stable country. The shortlist was between Durban and Mauritius. The Durban port was less efficient and had problems. There is exceptional logistical expertise in Mauritius, and it was also thanks to that that Mauritius was chosen. (MFD IT officer, January 2022)

To carry out this exercise in capturing goods flows, optimisation is essential, which explains the importance technological development has taken on within logistics since the beginning of the twenty-first century. After having developed its warehousing and distribution activities in the free port zone over the course of 15 years, beginning in the early 2010s, MFD undertook a major project to standardise its operations to simplify procedures and better structure its reception, storage, and distribution activities – all in the pursuit of organisational efficiency. As explained by one manager: 'We had hundreds of different processes; today we have 22. We changed our approach, we standardised our activities'. One important aspect of this renovation work was to acquire technology for the digital management of logistical operations, turning logistics into information work:

When you say 'logistics', people think of containers, transportation. What we do today is handle information. The actions that come at the end of a sequence—lifting up a box and putting it on a pallet—that's the easy part. It's everything that comes before, managing the information, that's the real heart of a logistician's job. And that job is tightly connected to information technology. Logistics requires automation and robotics solutions, and that attracts people from the IT sector we didn't have before. (Cybernaptics business development manager, January 2022)

In line with the findings of Posner (2018), digital technology is recognised as an essential tool for the modernisation and efficiency of the logistics sector, as one employee pointed out to us: 'Today, companies like MFD cannot function without digitalisation' (Cybernaptics Business Development Manager, January 2022).

This is where the company Cybernaptics comes in. Its founder, Viv Padayachy, was originally an employee at MFD, where he was head of information operations and services. However, in 2009, the company largely outsourced its IT operations, leaving Padayachy in a position to start his own business, which originally had six employees. One of them recalls: 'There were six of us in the beginning, and our first client was MFD, which gave us the support contract for their IT infrastructure, which we still have to this day'. From the outset, the company branded itself as an 'IT services' and 'business solutions' company, while also making clear its desire to position itself within the domain of 'software development'. By the end of the 2010s, Cybernaptics had approximately 50 employees and had become one of the most expert companies in the information/logistics nexus in Mauritius, as one of its department heads emphasises: 'Cybernaptics is the only Mauritian digital company that has expertise in logistics. We are also among the first to begin robotisation'. The heart of Cybernaptics' work is logistics, but the company offers various types of services, which can be divided into three main categories: hardware, software, and clouding. The company also intends to develop services in the area of the Internet of Things.

Cybernaptics' growth demonstrates the importance that digital technology has taken on in the logistics sector. As an employee, who had been working at MFD before joining Cybernaptics, emphasises:

> Before, everything was done by hand on paper; there were pre-defined zones and notebooks where the supervisors would write down in what zones the products should go. There were tons of paper at the end of the day. All that was digitalised with the use of an internal Wi-Fi network, Wi-Fi hand-held scanners; they go through the aisles and gather the products and get them ready for delivery. We reduced our paper usage because we print just the bar codes. (Cybernaptics sales engineer, January 2022)

Indeed, this transformation wrought profound changes, including more QR codes to allow for more precise scanning. Another employee mentions the use of RFID chips in bins of tuna, which facilitate and accelerate inventory production by allowing metal detectors to count the bins and make quantities visible in real time. He goes on to mention the fantasy of a warehouse without employees, where robotic carts go along the aisles and drones take from the inventory. But for Mauritian companies, such fantasies raise problems of cost that remain difficult to overcome: 'Everyone would like to have these technologies, but can we manage them, do we have the means to outfit a 100,000-euro warehouse in order to program drone inventory technology? Software and maintenance will cost us money' (Cybernaptics General Manager, January 2022). And yet, these words highlight the fundamental importance of digital technologies in running warehouses. In other words, it is possible to speak of a digital or information turn in Mauritian logistics. As Padayachy puts it, 'In the 2000s, when we talked about logistics, it referred to warehousing and transport. Today, logistics is much more of a way of managing information.' Thus, we shall now turn to the digital infrastructure that supports logistics.

THE DIGITAL INFRASTRUCTURE OF SEAMLESSNESS

One way to grasp the challenges faced by actors in operational capitalism is to identify the various tasks an organisation must oversee, from the arrival of a container at Port Louis to the reshipment of products by MFD. Our aim in doing so is to arrive at a kind of synthetic clarity, and we limit ourselves to the major tasks described to us throughout our study.

Logistics most often entails the intermodal management of merchandise transport. Goods are carried from their sites of production to sites of consumption through a combination of means of transport, including road, rail, maritime, and air networks. As one of the heads of MFD explains:

> We are laser-focused on the departures and arrivals of the ships. Because of how few ships are in circulation, when one arrives there is a great deal to unload, so it comes in waves. We have many containers that have to be

physically gathered and unloaded; it could be 50 to 80 containers at once to be gathered very quickly and unloaded. When you have a ship full of bicycles, it's difficult. In terms of volume it ends up being the same but it's in bits and pieces. It's not the same to have 3,000 containers that you can unload in 20 passes versus in 200. (MFD IT officer, January 2022)

In the case of unloading containers, there is a standard message for creating electronic reports between two commercial partners in order to certify delivery of a container: the 'container gate-in/gate-out message', or CODECO. The CODECO is part of a larger set of messages defined and standardised by the United Nations in 1987, the UN EDIFACT (UN Nations Rules for Electronic Data Interchange for Administration, Commerce and Transport). These standards are intended to structure datasets in order to facilitate exchanges between autonomous information and information management systems, a process already analysed in the case of the logistics of mining in the Central African Copperbelt (Blaszkiewicz 2023). In the case of MFD, the company receives delivery information from some of its main clients by way of a CODECO message. As one of the heads of Cybernaptics explains, 'CODECO is a messaging standard used internationally for container gate in and gate out in. We added a script so that the MFD system can generate CODECO messages to its client. MFD is using this with three of the biggest shipping lines in Mauritius'. However, two things must be noted here. First, the data structuring offered by the EDIFACT Standard does not guarantee an immediate connection between MFD and its client (for example, the shipping company Maersk). And second, since EDIFACT is but one set of standards among others, with some clients, different operations must be used to communicate between their information systems and those of MFD. These translation operations, which are indispensable for tracking ships and containers, are carried out by information solutions developed by Cybernaptics, as one of its company heads explained to us: 'The messages that clients send are all in different formats. We have scripts that monitor these messages and push them into the different systems. Reading these messages and pushing them into the systems requires specific applications'.

Once a container has been received, MFD staff unload it, palletise those goods that require it, and send the goods on to the warehouses. MFD runs warehouses that total approximately 70,000 square feet. Some of these warehouses are specialised according to product type (seafood and frozen goods that require specific temperatures; bulk goods; dry goods). While warehouses are essential elements in the logistical infrastructure provided by MFD, the importance of the digital systems required for storage – most significantly, the warehouse management system – must not be overlooked. As we saw in the previous section, in the middle of the 2010s, MFD undertook a significant renovation of its information systems. It decided to use the IT service management company Hardis, which specialises in warehouse management services. A warehouse management system makes it possible to identify and oversee goods, quantities, and locations in a warehouse. Software management is not the only service offered by Hardis, and as one of our interviewees emphasised, MFD chose the company because it also offers complementary services such as the installation of monitoring drones that automate the inventory. This inventory process, which covers everything from the set of pallets in a container down to the very object inside a box, involves creating and printing identifiers that are placed on merchandise units – a process that today is fully digitised.

Whereas up until the 2010s, software entry was done by hand, over the past few years MFD has acquired a fleet of hand scanners. Identifiers are paired with bar codes that warehouse employees can scan, on pallets, boxes, and goods. These scanners were provided by a leading company in the market: Zebra. This company, an American multinational, is one of the largest global suppliers of professional mobile handsets and solutions for data capturing, barcode printing, and radio frequency identification. With more than 8,800 employees and a client base in over 170 countries, the group provides tools that are indispensable for digitisation in the logistics sector, including mobile handsets for warehouse staff that range from pocket computers to vehicle-mounted computers and tablets with integrated digital intelligence, as well as scanners that can scan anything, anywhere, and in any conditions – including dirty and damaged bar codes – and can hold up in difficult work environments such as megaports. Cybernaptics was the first company to implement Zebra's solutions in Mauritius, and, as one of its managers explained to us, this know-how has also helped them develop informational and logistical activities in the rest of Africa: 'We were pioneers in offering Zebra's services in Mauritius. The knowledge we have of the port of Mauritius and of tools like Zebra—there are not many in Africa who have that. And we have been doing this for a decade. We are positioning ourselves as champions of trackability and mobility in the zone. We offer both software and hardware solutions' (Cybernaptics senior business developer, January 2022).

In order to make these scanners, printers, and computers work, warehouses needed to be fully connected to Wi-Fi, and this obligatory connectivity required specific adaptations and interventions. As Padayachy explained: 'Warehouses form metal cages that prevent signals from passing through. The metal structure creates shadow zones. Racks had to be connected with copper wires to prevent these shadow zones' (Cybernaptics general manager, January 2022). Warehouses are not only storage sites; they are also sites of constant information transmission. As one Cybernaptics employee explains, clients now have the option to view all their merchandise in the port zone: '[This happens via the] CFS, container fret system. You will have the number of clients for a specific container and the number of packages and their position in the warehouse. It generates gate passes to get your packages. You can have multiple containers and know exactly where your different packages are in different warehouses' (Cybernaptics software developer, January 2022).

These data collected in real time, stored, and transmitted by connected computers allow for a more efficient division of merchandise lots: once the contents of containers have been assigned to warehouses, new lots of merchandise are composed to be sent to various clients. Since there is software that allows for very precise subdivision of the contents of containers, even within boxes, clients can order specific quantities of products, down to the unit. From this point of view, one of the distinguishing characteristics of the warehouses managed by MFD – which serve not only clients across the Indian Ocean but also Mauritian clients – is that within them, merchandise lots are subdivided into very small quantities. And because some goods (for example, tubes of shaving cream) are delivered to small-scale Mauritian retailers, MFD has chosen to negotiate in very small quantities – which is sometimes a challenge, since some brands negotiate exclusive contracts with logistics companies – that is, if a logistics company distributes a company's products, it must supply all businesses in the zone, from supermarkets to small country stores. This precise and detailed handling of quantities led MFD to develop (and then to extend to all its clients) an online tracking system that constitutes an additional element of warehouses' digital infrastructure.

Once goods have been ordered by clients, the warehouse must then send them out. Warehouse workers load up goods, which are then transported in another container or by truck to a retailer on the island. Once packages have been sent out, another piece of software structures the transport activity: transport management software, which organises the allocation of routes, transport times, and the order of delivery. Of course, this is not the only digital tool used at this stage: for example, 'weighbridges' are used to measure the weights of trucks and to monitor any loss or diversion of goods during delivery operations. Developers at Cybernaptics describe different essential software they deployed on the port zone during recent years:

> Salesflex is a sales and distribution system. You have all your sales orders, your products, your clients. The salesmen use it to print receipts instantly, with a Zebra mobile printer. You can also manage payment through the service. You know exactly what products are left in your truck, what you sold. You have reports at the end of the day, by products, by clients, by type of payments. (Cybernaptics business development manager, January 2022)
>
> The online tracking system is a portal to generate reports and requests, to view and place orders, through a dashboard where clients can see delivery, reception, and stock reports and they can place new orders. It serves as a basic reporting tool on received goods, current stocks, and deliveries. The purpose is to minimise format errors in the system. The client sees all the reception, delivery, stock reports and they can place orders by selecting the articles and view their order history and delivery status. It's online, accessible. (Cybernaptics senior business developer, January 2022)

From CODECO messages providing data when containers arrive, to the delivery of half a dozen tubes of shaving cream to a small Mauritian retailer, we now have an overview of the multitude of software currently in use to enable all these logistical operations. In order to make this process 'seamless', many IT developers must work to allow these information suites to communicate between one another, since they each speak different languages. This patching activity, this work of creating patches, constitutes the heart of Cybernaptics' activity and the heart of the information turn in logistics.

PATCHING AND PATCHWORK

In this section, we argue that an essential function of digital activity within logistics is to allow for communication between the countless software tools that make up the digital infrastructure of logistics. The sociotechnological production of seamless and homogeneous flows involves ongoing work to bring tools into communication – an activity we refer to here as 'patching' and 'patchwork' – to underline alternately patching activity as it is being done, and patching as a category of labour. A better account and description of this phenomenon is necessary, first, because it makes it possible to better understand the materiality of the activity that goes into manufacturing digital infrastructures (the social production of infrastructures), and second, because it makes it possible to nuance the opposition between the supposedly innovative activity that takes place in the global North (software development) and the supposedly less innovative activity of maintenance and repair based in the global South (IT services). To make this argument, we draw on various works that have already pointed to the importance of connecting heterogenous infrastructures (Vertesi 2014) and of issues of interoperability within the digital environment of contemporary logistics (Gregson et al. 2017).

The interoperability of digital services – that is, their capacity to communicate between one another – is an essential characteristic of contemporary capitalism and its growing need for interconnections. It is a matter of an infrastructural process that must be ensured and constantly renewed, and which, once in place, is often forgotten – until a malfunction reminds us of its existence. This complex operation involves coding databases, languages, software, and equipment, and it has both economic aspects (for example, competition and profitability) and legal ones (for example, data protection). Indeed, one of Cybernaptics' heads of department states that for economic reasons, it does not develop its products from scratch:

> At the launch of Cybernaptics, we were developing from scratch. We saw that that wasn't profitable. Developers move around a lot, and we're not a huge corporation like Accenture, with hundreds of developers—if in the middle of development, a developer quits, it's hard. We focus more on customisations, integrations. We offer clients solutions that we currently have, and we try to customise the solution to meet the client's need. We have solutions that we have developed such as online tracking; that solution will go on top of their system. (Cybernaptics business development manager, January 2022)

The work of development is, however, not entirely absent from the company's activities, but it must be understood in terms of adaptation, summarised as follows: 'We first get an understanding of the client's needs and then propose systems that will match this need at the best price' (Cybernaptics business development manager, January 2022).

What are the major characteristics of this work of adaptation? We must first take into account the multiplicity of systems and the difficulty of making them interoperable, as one of the heads of Cybernaptics explains: 'Each company has its own software; we have developed EDI so that systems can communicate between each other. Doing so manually is not possible'. This is where Cybernaptics plays a front-line role, by interconnecting different types of databases and software. Transportation management and warehousing software are paired with other software that must constantly be integrated, as the following overview of Cybernaptics' various activities suggests:

> Another solution that is added on top of clients' systems is the CPMS, the Container Park Management System. For containers that are kept at the port, we can know which of those are going to leave first, according to how

> they are arranged. We also have a TMS, Transport Management System, software, that's for planning transport in term of the outflow and inflow of containers. This solution is integrated into the warehouses' management system (Reflex) as well as the accounting system (Navision). (Cybernaptics business development manager, January 2022).

This continuous adaptation of the software in place, its development and inter-communication, takes time and requires complex and individualised development projects lasting several months, as this developer explained to us:

> We specifically design solutions for specific clients. It can take us 2 to 3 months to develop and deploy a solution. From there we enhance it and adapt it to other clients. It's not the same system, it's different implementations every time and so it needs some development; for instance for Tag it took us 6 months to adapt the initial solution from one client to the next one. (Cybernaptics senior business developer, January 2022)

As an extension of this activity of interoperability, Cybernaptics also offers 'on top' services, which connect into the general software architecture to simplify clients' processes:

> We developed Salesflex, which is a mobile vending tool, an Android app that can be connected to any software like Navision or Oracle, and if you have mobile vendors, they can sell directly from their mobile device where they have their stock downloaded, make sales, manage payment and invoices, and they can even push sales data directly into their accounting systems. Before, they would have a notebook where they wrote down sales and had to enter them into the system later at the office; now they can do it immediately. (Cybernaptics senior business developer, January 2022)

'On top' services are developed on the basis of the software suite already in place; they add a function or service that the original software (which is often proprietary) either does not include or for which the cost is too high.

The question of cost is absolutely crucial to understanding the context in which patching is mobilised. This activity takes place in a particularly low-resource context, as one of the Mauritian developers we met explained: 'In Africa, there are many clients who cannot afford Zebra. But with lower-cost products, they can move technologically and digitise their operations.' Patching is also used to pay less or to be able to operate with degraded equipment, since proprietary materials and software are financially, and sometimes geographically, inaccessible (in Africa, for example, or in the middle of the Indian Ocean). As explained in the same interview:

> In terms of price, Zebra is not at all accessible to everyone; it's like a Tesla or a Porsche. Today, there are non-branded products from China that penetrate the market. Whereas a Zebra scanner costs 2,000 euros, these can be bought for less than 500. To respond to this competition from China, we have a product like PointMobile, which is from Korea and meets our needs for quality at a lower cost than Zebra. Initially, we tested Chinese non-branded products but the quality was lacking. (Cybernaptics senior business developer, January 2022)

The business has adapted to these cost challenges and offers software customisations from an Android operating system, in order to provide lower-cost software solutions to its African clients who have software needs but limited budgets.

The issue of technological innovation constitutes another important challenge for enterprises like Cybernaptics. The engineers we met in Mauritius highlight two innovations that today have become indispensable, and which have major consequences for their work and for the dependencies and costs associated with the digital infrastructures of logistics: clouding and automation.

The current use of 'clouding' services raises numerous security problems: 'Before, we had in-house physical servers. Now, with the cloud, there are new services to securitise cloud servers; firewalls and internet access control are important'. These developments took place very rapidly, in less than 10 years. The adoption of the cloud has now become routine, as explained during the interview: 'Cloud hosting has become very popular, and as a result we have fewer physical servers'. Various actors within the Mauritian IT sector mentioned the challenges of data security and dependency on cloud service providers such as Amazon and Google. What are the risks involved in shifting from a company's physical servers to Amazon servers? Some logistics actors prefer to maintain their own physical servers for security reasons as well as for reasons of data sovereignty, but the COVID-19 crisis dramatically accelerated the clouding of business data, which is increasingly hosted by these large-scale private American operators. The new technological challenge for Mauritian IT workers is to ensure data security at a distance, using new firewall and blockchain technologies to encode and protect as much as possible the data circulating from one cloud to another. However, there are few IT workers on the island trained in these technologies.

The trend toward automation constitutes another imperative for technological innovation that Mauritian operators both implement and question simultaneously, as one manager explained with regard to a project to automate exchange rates: 'The trend is robotisation, but robotise what? We see that the demand is coming from the financial sector but we weren't familiar with that field at first. So we developed solutions alongside banking sector experts. That solution isn't going to work for all the actors in the sector'. In this context, Cybernaptics also offers solutions that make it possible to connect heterogenous environments automatically. Our interlocutors also mentioned the automation of warehouse inventories by drones, explaining that such systems come with considerable costs that would need to be amortised, thus raising the question of whether it is profitable to automate inventories of low-cost basic items such as potatoes or flour.

All these various tasks require patching and thus presuppose the intervention of actors such as Cybernaptics, which allow various environments and databases to communicate between one another. This leads to the question: What is 'customisation' in these cases? What does a 'tailored' solution mean, and what does the adaptation or matching of a solution to a consumer involve? The answers to these questions are particularly unclear given that the explanations offered by Cybernaptics and MFD employees often contain veritable contradictions: for example, when one declares that the customisation of systems is what is most important, and that this is what Cybernaptics does, whereas, for another employee, the most important is the service aspect of their work: 'We do customisation but also support; that's very important. It's the service that comes with the software that is the most important. For us, it's not just about selling the solution.'

Here, we emphasise the importance of interoperability, which Pelizza (2016) has analysed as a 'process of institutional reordering' (Pelizza 2016: 305). Interoperability requires several conditions. There must be datasets: sensors and forms must have been 'populated', which requires interfaces. Datasets must be accessible: they must be stored and circulated on networks, whether they are created ad hoc, requested, or purchased. They must be clean and, above all, interoperable. They must 'speak' the same language, so that they can be connected via a common frame of reference. The goals of interoperability are defined and inscribed in the specifications given to developers. These determine the model of data mining that developers create, going back and forth between development and testing until the moment of transfer to the production environment. Lastly, interconnections operate within a social organisation that determines the possible use of their results. All of these elements are sociotechnical: none is a given, and each raises questions of regulation, depends on the circulation of data, and requires the coordination of expertise and interests, data providers, workers, and users.

CONCLUSION

This case study contributes to show how seamlessness is manufactured on the ground. Although it does not document other aspects of that production, such as the warehouse or ship labour analysed by other authors (Flécher 2023; Gaborieau 2016), it illustrates an important characteristic of contemporary capitalism: the integration, within a single site, of logistical, digital, and financial expertise and infrastructure. The Mauritian model of development perfectly illustrates the regeneration of capitalism through strategies for attracting flows of goods, which involve the increasing integration of the logistics-finance-digital-technology triad. Thus, it becomes important to turn toward more nuanced analyses of the digital infrastructure that supports the logistical activity essential to the international circulation of flows of data and goods.

We have endeavoured to show the central role of digital tools in this activity. From communications between container carriers to their arrival at the free port and the unloading of pallets, all the way through to delivery to retailers on the island, logistical operations depend on a multitude of software programs. To make this movement of merchandise 'seamless', many IT developers must work to make information programs that speak different languages communicate between one another. However, the human and financial resources necessary to implement this interoperability, which is now indispensable, are lacking in developing countries.

African countries, following Mauritius's lead, are looking for more economical and accessible solutions in order to meet the infrastructure and digital challenges that today are central to the circulation of merchandise flows. From this point of view, we may say that Cybernaptics embodies a trend toward frugal, efficient, tailored seamlessness, and reminds us of the degree to which intermediaries play an indispensable role in the mobilisation of digital environments and in the creation of innovation under constraints, with limited human and financial resources. This activity of adaptation, patching and patchwork constitute the heart of the information turn in logistics observed from the point of view of global Southerners. It therefore highlights one specificity of technological globalisation in Global South countries: seamlessness production not only answers to a requirement set by global logistical capitalism but is also a strategy through which social actors make commensurable IT systems or technological tools that largely differ in nature. The IT communication and track and trace device standards imposed by the most powerful companies and countries cannot be followed by the vast majority of commercial actors in Global South countries, due to the lack of resources and to their frequent inadequacy to local contexts. As a consequence, the IT systems and devices coexisting in hubbing locations such as Mauritius are highly heterogeneous.

In a way, patching is not specific to Global South countries: coding and manufacturing IT systems are patching activities at the most basic level. The singularity of Global South countries is, however, to be continuously exposed to this activity to cope with systems promoted – sometimes imposed – by the most powerful players of the industry and mostly financially inaccessible for small-scale firms operating in the South. In this sense, Global South patching presents specificities as compared to patching 'in general'; it also gives indications of what patching could be even in richer settings, as some of them go through crises and become weakened. Patching from the South then appears as a way for actors to overturn financial and infrastructural constraints by offering adjustment solutions between heterogeneous systems. In that sense, patching is playing a key role in globalisation processes by making heterogeneous systems, often determined by differences in financial power, commensurable and interoperable. It is simultaneously an answer to the seamlessness imperative and a situated strategy to circumvent and subvert situations determined by the globally uneven distribution of resources and technological equipment – and in turn it could also become a model for flow capture and global flow manufacturing without regard for location, be it in the North or in the South.

REFERENCES

Blaszkiewicz, H., *Toujours plus vite? Logistique et capitalisme dans l'Afrique minière – Zambie, RD Congo* (Paris: Le Manuscrit, 2023).

Chazan-Gillig, S., and I. Widmer, 'Circulation migratoire et délocalisations industrielles à l'île Maurice', *Sociétés contemporaines*, 43 (2001): 81–122.

Cowen, D., *The Deadly Life of Logistics: Mapping Violence in Global Trade* (University of Minnesota Press, 2014).

Flécher, C., *À bord des géants des mers: Ethnographie embarquée de la logistique globalisée* (La Découverte, 2023).

Gaborieau, D., 'Des usines à colis: Trajectoire ouvrière des entrepôts de la grande distribution', (PhD Thesis, Université Paris I, 2016).

Grégoire, E., 'État développeur, État fragile: Comment l'île Maurice est-elle devenue un pays émergent alors que le Niger demeure un PMA?', *Autrepart*, 80 (2016): 3–23.

Gregson, N., M. Crang, and C. N. Antonopoulos, 'Holding Together Logistical Worlds: Friction, Seams and Circulation in the Emerging "Global Warehouse"', *Environment and Planning D: Society and Space*, 35 (2017): 381–98.

Mezzadra, S., and B. Neilson, *The Politics of Operations: Excavating Contemporary Capitalism* (Duke University Press, 2019).

Neveling, P., 'Three Shades of Embeddedness: State Capitalism as the Informal Economy, Emic Notions of the Anti-Market, and Counterfeit Garments in the Mauritian Export Processing Zone', *Research in Economic Anthropology*, 34 (2014): 65–94.

Pelizza, A., 'Developing the Vectorial Glance: Infrastructural Inversion for the New Agenda on Government Information Systems', *Science, Technology, & Human Values*, 41 (2016): 298–321.

Posner, M., 'See No Evil', *Logic(s) Magazine*, April 2018 https://logicmag.io/scale/see-no-evil/.

Ramtohul, R., and H. T. Eriksen, *The Mauritian Paradox: Fifty Years of Development, Diversity and Democracy* (University of Mauritius Press, 2018).

Rogerson, C. M., 'Export-Processing Industrialisation in Mauritius: The Lessons of Success', *Development Southern Africa*, 10 (1993): 177–97.

Schnepel, B., 'The Making of a Hub Society: Mauritius' Path from Port of Call to Cyber Island', in B. Schnepel and E. A. Alpers, eds., *Connectivity in Motion: Island Hubs in the Indian Ocean World* (Palgrave Macmillan, 2018), pp. 231–58.

Vertesi, J., 'Seamful Spaces: Heterogeneous Infrastructures in Interaction', *Science, Technology and Human Values*, 39 (2014): 264–84.

8

HALTING THE 'FORCED MARCH': THE UPS AND DOWNS OF CHAD'S INTEGRATION INTO GLOBAL PHARMACEUTICAL MARKETS

Ilyass Mahamat Nour Moussa

ON TUESDAY, 3 SEPTEMBER 2019, THE CHADIAN PHARMACEUTICAL MARKET found itself with a new World Bank–funded laboratory for the analysis of the quality of industrial pharmaceuticals. Its inauguration was a solemn event attended by Chadian officials and representatives of the World Bank at the premises of the Centre de contrôle de qualité des denrées alimentaires (CECOQDA) – the French expression designating the Centre for Food Safety – where the laboratory was established. The Chadian press showed up to cover the ceremony, which was broadcast on national television to introduce the population to the facility's various technological analysis instruments. Hissein Tahir Sougoumi, Chad's secretary of state for the economy and development planning, declared, 'Chad has just been equipped with strong, independent capabilities in the fight against falsified medicines'. He justified the increase in medicine control by the weakness of the pharmaceutical supply system.

The mention of 'falsified medicines' in his speech was significant of the government's efforts to act on the recommendations of international organisations to secure pharmaceutical markets. In this regard, the creation of the laboratory was more generally indicative of the government's alignment with a global model of effective pharmaceutical management, promoted by international organisations such as the World Bank, and materialised by the deployment of regulatory and safety mechanisms. At first sight, this model could be described as an ongoing effort to consolidate and strengthen pharmaceutical supply chains at global and national scales. But the meaning of 'consolidation' and 'strengthening' should not be taken for granted, as securing the circulation of medicines in the Global South has also been identified as a key strategy of Northern companies and countries to assert their hegemony and dominance over global pharmaceutical markets (Baxerres 2015; Hornberger 2018; Quet 2022).

In this chapter, I discuss these complex articulations between the global models, their supposed imposition, and their local implementation. I show that the implementation of Chad's laboratory is not simply the replication of a dominant model of pharmaceutical control in order to comply with global norms. To do so, I build upon a vast literature dedicated to the social study of pharmaceuticals in Global South countries. Some of this literature has analysed in depth the weight of control and regulatory operations upon medicines in pharmaceutical markets as they pass through various stages (Desclaux and Lévy 2003; van der Geest 2017; van der Geest and Whyte 2003; van der Geest et al. 1996). It has also offered a different reading to the circulation of medicines within the healthcare system, based upon several studies in Sub-Saharan settings (Desclaux and Egrot 2015; Fassin 1986, 2000; Peterson 2014; Whyte 1992; Whyte et al. 2002). In continuation with this scholarship, I highlight some of the ways in which actors located in a Global South country appropriate and circumvent a system designed to secure the flow of medicines (Quet et al. 2018). I demonstrate that the integration of Chad in global pharmaceutical markets is largely driven by power asymmetries between the North and the South. But I also show that if Southern governments are generally poorer, need more external financing, and are therefore more dependent on outside sources than their Northern counterparts, this does not mean that the financial flows that come their way will force them to do what funders want. The social actors, in governments as well as in the private sector, understand the rules of the game and know that they need to respect them to keep playing, but this does not necessarily mean that they will follow them blindly.

Hence, I want to nuance the idea that the 'forced march' initiated by the standardisation of regulatory work on the Chadian pharmaceutical market is solely a top-down process where local actors do not have room for manoeuvre (Radnóti and Moreau 2000; Rawicz and Chédaille 2011). Analysing how technical tools and regulatory frameworks are implemented in Chad through the case study of the creation of a national control laboratory helps us question the idea that external funders are omnipotent in shaping pharmaceutical markets and medicine flows in one country. This case will be helpful for thinking about and reflecting on how Global South players are anticipating, but also appropriating, repurposing, and circumventing, international organisations' recommendations in order to fit their own needs.

I draw on empirical data collected during various field missions carried out in N'Djamena (Chad) between 2020 and 2023. During eight months of fieldwork, I held 50 semi-structured interviews with individuals directly involved in the creation and implementation of the laboratory, as well as members of international organisations, the pharmaceutical regulatory authority, and wholesale distributors. I also carried out direct participant observation at the medicine quality control laboratory in N'Djamena. As the laboratory was closed to the public, all my initial attempts to obtain information about its funding and its actual or supposed operation failed. Later, I got access to the laboratory through an acquaintance who worked as the head of food quality control at CECOQDA. Thanks to him, I could contact the manager and the seven other technicians of the laboratory. Throughout my meetings in the laboratory, I aimed to observe the actual practices of the technicians and understand how technical systems were set up for drug control operations. During this immersion in the laboratory, I created a trusting environment with my interlocutors by listening to rather than intensively questioning them, which helped me capture nuanced information that might not be obvious through more directive or structured enquiry methods.

The chapter is structured as follows. In the first section, I show how international organisations have played and continue to play an essential role in the global integration of the Chadian pharmaceutical market by supporting a specific approach to the securitisation of medicine flows. However, the Chadian state is far from being a passive recipient of global funding flows, which I show in the second section based upon the presentation of the SWEDD project – a project that local actors have diverted to suit specific needs as much as they have answered to external demand. Sections three to five then detail my argument regarding local adaptation by explaining how the pharmaceutical control tools are circumvented, diverted, and appropriated by the lab technicians, reflecting some of the dynamics of technoscientific globalisation 'from below'.

INTERNATIONAL ORGANISATIONS AND THE SECURITISATION OF PHARMACEUTICAL MARKETS IN CHAD

Chad has been engaged for several years in (re)building its national pharmaceuticals market, with the support of international organisations such as the World Health Organisation and the World Bank, among others. The World Health Organisation has played a primary role in the reframing of Chadian pharmaceutical markets, notably through regional initiatives. In May 2005, it funded the organisation of a workshop dedicated to the creation of a regional pharmaceutical policy that gathered the representatives of Central African Economic and Monetary Community (CEMAC) pharmacies in Yaoundé (Cameroon).[1] This led to the elaboration of a pharmaceutical policy launched in 2014 and shared by the six member states (Cameroon, Central African Republic, Congo, Gabon, Equatorial Guinea, and Chad). The WHO acted as guarantor of this initiative to resolve dysfunctions in the various pharmaceutical systems by standardising the rules and functioning of pharmaceutical markets. In addition to these standardisation efforts, the organisation encouraged the regulation of markets at a regional scale. It organised training geared towards health centre pharmacists and journalists (on issues pertaining to pharmaceutical regulation), and contributed to setting up a network of journalists 'Amis de la pharmacie' (Pharmacy's friends) for the fight against substandard and falsified medical products. The organisation has also been advising the implementation of significant regulatory measures in the CEMAC area through multiple initiatives, to such a point that, according to a WHO representative, the Central African community has become a benchmark.

Though at a different scale, the World Bank has also been one of the main organisations contributing to the revamping of the Chadian pharmaceutical market. Its implication in the Chadian pharmaceutical field goes back to 1994, when the bank funded the setting up of the Centrale Pharmaceutique d'Achats, the pharmaceutical procurement agency. More recently, it funded in 2019 the national pharmaceutical quality control laboratory studied in these pages. This contribution to pharmaceutical development is part of the activity of the bank in Sub-Saharan Africa. During the same period, it financed a similar laboratory in Mauritania and several initiatives to secure the flow of medicines in East Africa. Since 2022, there have also been joint initiatives between the bank, the West African Health Organisation (WAHO), and the United Nations Population Fund (UNFPA) to set up a system for controlling the flow of cross-border medicines. This web of relations between health actors financed by the World Bank has led to the creation of a subregional network of National Medicines Quality Control Laboratories (NMQCLS) in the Economic Community of West African States (ECOWAS), and additionally in Mauritania and Chad. The UN agency (UNFPA) and the subregional institution (WAHO) not only created the network but have also been actively involved in training Chadian laboratory staff and are even planning to launch a programme to monitor the percentage of non-compliant and falsified medicines.

As shown above, although the WHO and the World Bank are the most visible players, the project of (re)construction of the Chadian pharmaceutical market has involved multiple other organisations and has been unfolding at various levels in the field, starting with the activation of a former project to build a medicine manufacturing plant, which is still underway; the plan to set up a pharmaceutical regulatory authority; the creation of a medicines quality control laboratory; the opening in 2011 of a pharmacy department at the University of N'Djamena; the revitalisation of the drug commission; the training of pharmaceutical inspectors; the establishment of a health police force; the imposing of medicine traceability on private players; and the organisation of regular meetings between health players on medicine-related issues.

Beyond their diversity, one important aspect of these organisations' engagement with Chadian pharmaceutical politics has been their emphasis upon speed (of change) and security. One example was given in 2019 by the organisations, via the Ministry of Health, of a national meeting on the topic of pharmaceuticals in N'Djamena, with exclusive funding from the WHO. The objective of the meeting was to advance an essential medicines policy and to align with the standards and guidelines recommended by the WHO in the field of pharmaceutical procurement and regulation. The WHO representative in N'Djamena asserted on this occasion that there was a significant discrepancy between the realities of the Chadian pharmaceutical market and the objectives of his institution. He pinpointed the slow pace of progress in the Chadian pharmaceutical market towards the WHO's objectives. In order to make up for this loss of time in terms of legislation and regulation, it was necessary to speed up the improvements. He also suggested that health professionals and the Chadian government should move forward the harmonisation of the legal framework to combat substandard and falsified medicines (MQIF), establish a medicines regulatory agency or authority and a health police force, accelerate the MEDICRIME membership process and set up a national medicines quality control laboratory. His speech, both emphasising the need for more pharmaceutical security and the importance of acting fast, was revealing of the priorities set up by the WHO for Chad.

How can this drive to secure pharmaceutical products on the Chadian market be explained? Part of the answer is found in an interview conducted during the summer of 2021 in N'Djamena with managers of Laborex Chad, one of the major wholesale distributors of the Chadian pharmaceuticals market.

> Some American companies refuse to sell us medicines because they feel that safety conditions in Chad are unreliable. By safety, we don't just mean quality control of pharmaceutical products, but a whole process that consists in protecting Chadian products and consumers against risks.

From this interview with my interlocutor, it appears that safety and security are strongly emphasised by pharmaceutical industries. To import industrial medicines into the pharmaceutical market, certain laboratories require specific guarantees from Chadian wholesale distributors. This requirement by some pharmaceutical companies is a product of market normativity. Since 2020, some wholesalers have been reorganising their operations and establishing warehouse units that meet pharmaceutical companies' criteria, such as maintaining a temperature below 25°C in the warehouses and developing a traceability system for the distribution of medicines on the local market. This sums up the vision of the external actors involved in this new reconstruction of the local pharmaceutical market. Refusal to comply with this injunction from the pharmaceutical industry could exclude a wholesaler-distributor, or at least interrupt their partnership with a pharmaceutical company.

Chadian stakeholders have described this framing of the pharmaceutical market regulatory work by international organisations as 'extreme regulation' and as a 'forced march' imposed on a country that has depended on medicine imports since it acceded to national sovereignty and where the pharmaceutical regulatory system is still characterised by many shortcomings. The expression 'forced march' refers to the various reforms of the pharmaceutical system that Chad has undertaken to comply with the requirements of international organisations, from the amendment of the pharmacy law to the creation of a medicines quality control laboratory and the requirement for pharmaceutical products to be traceable to wholesale distributors. The forced march is seen by multiple Chadian actors of the pharmaceutical sector as an obligation to drastically change practices, with no possibility of turning back, in order to achieve as fast as possible a goal defined externally. Through regulation at all costs, international organisations, with the support of a few national players, would seek to control industrial pharmaceutical circulation.

> Today, several technical and financial partners are intervening in the Chadian pharmaceuticals market through various sanitation projects, which increasingly resemble a drive to control the circulation of medicines. The measures used by Chad's partners and their discourse around the issue of safety seem somewhat exaggerated to me. (Interview with a senior health technician and manager of a pharmaceutical depot, N'Djamena, summer 2022)

This excerpt from an interview with the manager of a pharmaceutical depot in N'Djamena highlights the differences between the international organisation's framework and local actors' considerations regarding the security of industrial medicines in the Chadian pharmaceutical market. The rhetoric of security as put forward by the organisations is not universally accepted on the ground. Local actors even see it as a form of domination exerted by organisations such as the World Bank, the World Health Organisation, and other international institutions, all advocating for the securitisation of flows and, by extension, protecting the interests of large Northern pharmaceutical firms.

At first sight, it thus looks as if Chad's pharmaceutical policy is largely dictated by international institutions based on global standards and through the implementation of technical devices. However, I show in the following section that the implementation of these standards is filtered by the local actors, who have agency in using development funds and setting up global norms.

'DIVERTING' GLOBAL DEVELOPMENT FUNDS: THE ESTABLISHMENT OF THE LABORATORY

The complex articulation between the standardised models promoted by international organisations and the reality of practices can be better illustrated through the history of the national laboratory for quality control. Building a national laboratory for the quality control of industrial medicines requires funding, something the Chadian state mostly lacks. However, African states often host development programmes that are well endowed with finances, as in the case of the regional initiative for Sub-Saharan Africa Women's Empowerment and the Demographic Dividend (SWEDD), notably supported by the United Nations and the World Bank. The initiative was launched by the Sahel countries (Burkina Faso, Côte d'Ivoire, Mali, Niger, and Chad) in 2015 against the backdrop of multiple crises (humanitarian, security, and environmental) affecting this part of the African continent: weak state institutions, corruption, poor governance, and ethnic tensions exacerbated existing problems in the Sahel region. The combination of these multiple crises had initiated a vicious cycle of mutually reinforcing poverty, insecurity, political instability, and forced displacement. This was the context in which the SWEDD project was born and was subsequently supported by international institutions, first and foremost the World Bank. Through its International Development Association (IDA), the World Bank provided US$376 million, bringing the total funding to US$680 million for all the countries concerned with the project. This brings us to ask: given the scope of the SWEDD project, initially designed to support women in a context of multiple crises, how did the World Bank end up investing in the Chadian pharmaceutical market through it?

The SWEDD project, which was instrumental in setting up the medicines quality control laboratory, included a health arm to meet the demand for reproductive, maternal, neonatal, infant, and nutritional health products and services (SRMNIN). This arm mobilises significant resources on the ground, focusing on the health of women and infants. It develops programmes to improve access to maternal health care, such as prenatal consultations, skilled birth attendance, and postnatal care, and initiatives to promote family planning and reproductive health. The health arm of the SWEDD project also addresses issues such as sex education, early pregnancy prevention, sexually transmitted diseases, and HIV/AIDS. It incorporates nutrition-related interventions to combat malnutrition among women and children and to build the capacity of local communities to take charge of their health. This includes training community health workers, providing access to essential medicines, promoting hygiene, and supporting community-based disease surveillance systems. The health arm furthermore contributes to strengthening health infrastructures by renovating and extending them, particularly in the most remote and vulnerable regions. It is apparent from this description of the project that, although it did have a 'health' component from the outset, this did not include frameworks for securing the flow of medicines, which would subsequently receive World Bank support. The involvement of international institutions such as the World Bank has since been growing steadily within SWEDD, revealing a contradiction with the project's initial objective of combating poverty and empowering women.

Through its health infrastructure development arm, the project financed the establishment and equipment of a quality control laboratory for medicines under the supervision of the Pharmacy and Pharmacopoeia Department of the

FIG. 8.1 National coordination of the SWEDD project, through which the World Bank financed the installation of a quality control laboratory, N'Djamena, July 2022 (Ilyass Mahamat Nour Moussa)

Ministry of Public Health. Initially, the aim was to facilitate women's access to essential medicines. However, a subsequent shift was apparent in the project's objective towards the fight against falsified medicines, leading to a quality control laboratory. According to the person in charge of the health component of the SWEDD project, 'it was the Ministry of Public Health that brokered the project and expressed the need to equip a laboratory theoretically created by the Chadian state, but which lacks everything required for its effective operation'. While this comment seems to indicate that the laboratory project was born out of a need expressed by Chad, it is significant that such a need had been embedded in the global context of the world's pharmaceutical market for several years. We thus must nuance the idea that the Chadian state 'diverted' the funds; it did something not initially expected by their funding provider but nothing that went entirely against their interest and logic, as explained by the person in charge of the health component of the SWEDD project:

Today, there's enormous pressure exerted by pharmaceutical companies worldwide. This weighs on wholesalers in terms of securing pharmaceutical products. Several pharmaceutical companies don't want to trade with countries that don't set up an effective monitoring system, and so Chad wanted to comply with this so that it could continue importing medicines.

This quotation confirms that in addition to the health activity developed by the SWEDD project and the World Bank's financing of the laboratory, Chadian authorities have committed to securing the country's pharmaceutical supply to meet the requirements related to the globalisation of pharmaceutical markets.

By 'asking-agreeing' to set up an industrial medicine quality control laboratory on its territory, Chad was trying to respond to the WHO's request to control the flow of industrial medicines within its borders. Two factors in particular explain why the Chadian government mobilised in favour of this installation. The first concerns the knowledge produced regarding the Chadian market: several reports from the Ministry of Health indicate that the black market for medicines has a supply and distribution capacity ten times greater than the official market. The Chadian state thus seeks to limit potential crises related to the consumption of unauthorised medicines and emphasises a public health issue, in line with the recommendations of international organisations. This move is supported by local healthcare professionals and some wholesale distributors. For these distributors, the laboratory installation would slow down the unofficial trade of medicines and consequently provide an opportunity to sell more medicines in the pharmaceutical market. But the second factor is mostly financial: in setting up the national drug quality control laboratory, the Chadian state gave a sign of good will that allowed it to attract funding from World Bank and the WHO, in order to invest more broadly in other sectors. In that perspective my field observations show that the Chadian authorities not only answer to ideological and normative public health pressures, but they also navigate through external financial constraints in order to achieve their own goals. However, even though the industrial medicine quality control laboratory could now monitor the flows, local actors in the pharmaceutical system would follow rationales and interests that might hinder its proper functioning.

THE WEAK APPROPRIATION OF TECHNICAL TOOLS

Once it was set up, how was the laboratory appropriated by Chadian actors? To explain this phenomenon, we can resort to the notion of 'diversion' developed by Mathieu Quet (2018) in the case of hepatitis treatments, after Claire Beaudevin and Laurent Pordié (2016). In the Chadian context, the technical tools for drug quality control are somehow diverted by the laboratory technicians, the private actors, and the state. However, to push this notion further, I insist on the variety of practices through which diversion is manifested. In the following sections, I analyse the inner workings of the laboratory and show the difficulties in appropriating the technical equipment for controlling the quality of medicine to global standards, the hardship in convincing the distributors to get their product tested in the lab, and the repurposing of technical facilities for hydroalcoholic gels in times of COVID-19.

Installing technological equipment to control the quality of pharmaceutical products is part of a series of reforms carried out over the years to combat falsified products. It requires the use of several instruments pictured in Figure 8.2. For instance, HPLC (high-performance/pressure liquid chromatography) columns are used for the analytical and preparative separation of molecules in a mixture. The lab also hosts ultra-performance liquid chromatography (UPLC) instruments, which offer a slightly different level of analysis with good chromatographic resolution, and an ultraviolet-visible spectrometer (UVS) for breaking down a quantity observed in a light beam – for spectroscopy – or in a mixture of molecules. The laboratory is also equipped with disintegrators, analytical scales, and dissolution testers, enabling quality control of a drug through the substances used by the pharmaceutical industry. In this physicochemical laboratory, two types of control are carried out using these devices. The first is a physical drug analysis, which checks the packaging, expiration date, country of origin, package inserts, and other physical elements. The second check is an in-depth chemical analysis of the drugs. This quantitative and qualitative analysis checks the presence and composition of active ingredients.

All these devices should enable the laboratory's technicians to carry out quality control of medicines destined for the Chadian pharmaceutical market.

FIG. 8.2 Overview of the building housing the medicines quality control laboratory, N'Djamena, June 2022 (Ilyass Mahamat Nour Moussa)

According to the laboratory's quality manager, 'in principle, with such [devices], we can reduce our dependence and make consumers trustful'. The dependency highlighted by my interlocutor in this interview is linked to the use of drug quality control laboratories located abroad. Indeed, before the advent of the Chadian laboratory in 2019, importers of industrial medicines into the Chadian pharmaceutical market were required to have a sample of the drugs tested in countries such as Niger, Kenya, or France before receiving clearance from regulatory authorities, which allows for the distribution of the medicines in Chad. The difficulties in implementing projects to secure medicines in the Chadian pharmaceutical market are reflected by the various processes of appropriation by local actors.

The concept of appropriation is used in various fields. However, it generally refers to the process whereby a person or group assumes control of something, be it an object, an idea, a culture, or even a space, for its own purposes. This usually involves claiming ownership, control, or use of a good or an aspect of a culture and can have positive or negative consequences, depending on the context and how it unfolds on the ground. The sociology of use and innovation is helpful here. It has documented appropriation practices for years, implementing instrumental approaches through the internalisation or externalisation of technical objects (De Vaujany 2006; Morigi and Braga 2014). My own investigation leads to the observation that the appropriation of this technical facility by the local pharmaceutical players was first and foremost very limited, as testified by the manager of Laborex Chad.

> Today, in the Chadian pharmaceutical market, the Centrale Pharmaceutique d'Achat is the only wholesaler to have its drugs checked by the national medicines quality control laboratory that's been created.

Since it was set up in 2019, the Centrale Pharmaceutique d'Achats (CPA), as a public wholesale procurement agency, has had its medicines controlled by the laboratory. To date, it is the only Chadian wholesaler involved in distributing industrial pharmaceuticals on the market to have had its imported industrial pharmaceuticals inspected – out of over 38. The CPA's use of the quality control laboratory is explained by the nature of the institution – financed by the World Bank – and by its mission to ensure the supply of medicines for the public sector. At the turn of the 1990s, the World Bank financed the establishment of drug import institutions in Africa, for example in Benin, where the Central Purchasing Center for Essential Medicines and Medical Consumables (CAME) began operations in 1991 (Mahamé and Baxerres 2015). This period highlights the World Bank's paradigm shift to focus on poverty reduction in developing countries and on building infrastructure. In keeping with the promises made by the Chadian government to international organisations, the CPA presents a semblance of normativity in relation to the purposes announced during the laboratory's inauguration in 2019. Thus, official discourse and the laboratory's actual use are in contradiction, highlighting the tension between global objectives and local needs and practices.

THE CIRCUMVENTION OF QUALITY CONTROL

If the CPA imports solely generic drugs, the wholesale distributors import branded or specialised drugs. For several of these players, the analysis of brand-name drugs may require a level of experience not available to the technicians of the Chadian laboratory. The apparent struggle of the Chadian technicians in appropriating some technical tools was often used as a reason for circumventing the drug control process at the laboratory established in N'Djamena. As with appropriation, the circumvention principle captures another strategy players resort to in the Chadian market. It involves the non-take-up of technical facilities provided by the World Bank, such as the medicines quality control laboratory. The Chadian pharmaceutical market includes 38 private wholesale distributors (alongside the publicly funded CPA) that take part in the distribution of medicines but try to bypass the laboratory for the quality control of imported pharmaceutical products. They have never resorted to the services offered by this laboratory, in spite of the stipulation in Chadian pharmaceutical regulations that quality control of imported medicines is mandatory, whether by the national system or approved foreign services:

> Pharmaceutical products imported for the Chadian market have to pass through Niger, Kenya, or France for certification before being distributed on the local market. This means that the products have to be quarantined, and as soon as the wholesaler shows us the certification of its merchandise, the authorities can lift the quarantine on these products, but unfortunately, the opposite is happening. (Interview with the director of pharmacy and laboratory, N'Djamena, 2021)

This excerpt from an interview with the director of pharmacy shows that wholesalers circumvent the system by distributing medicines without first having them tested by the quality control laboratory. How can such circumvention be explained when the system was created precisely for these players? There are several reasons. First, analysing samples in this laboratory means quarantining the products during the analysis period, whereas wholesale distributors look for quick profit from imports. One, based in N'Djamena, explained: 'Despite all the talk of quality control in this new laboratory, the waiting time can be a bit long, so we prefer to bypass the system and not quarantine the medicines'. While time is one of the factors put forward as a reason for circumventing the laboratory, the atypical profile of wholesale distributors also plays a vital role in this choice. A large proportion of them are not pharmacists but traders, and the few pharmacists involved in importation are associates or nominees, as explained by the head of Laborex Chad:

> More than half of Chad's wholesalers are not pharmacists. The country does not have enough pharmacists, and the few who do work in the Chadian market don't have the necessary means to run a wholesale business in the current context. (Interview with head of Laborex Chad, N'Djamena, 2021)

In addition to the lack of laboratory technicians' experience, the importers' profiles also play a significant role in circumvention. Among the 38 wholesale distributors, fewer than 10% are pharmacists associated with the companies importing industrial medicines. This low percentage of pharmacists in the importing companies may facilitate circumvention. At this level, circumvention refers to the refusal to quarantine imports before distribution in the Chadian pharmaceutical market. Additionally, the failure of wholesalers to comply with quarantine requirements has been facilitated by the corruption that has plagued the sector for several years. The close relations between certain Chadian customs officials and wholesale distributors facilitate this corruption and non-compliance with drug quarantine, which benefits private players. Customs is, moreover, not the only factor explaining corruption within the pharmaceutical system; community ties also play an essential role in circumventing technical facilities.

> It is true that some customs officials are very helpful, but not everyone. Several senior customs officials are part of the Zaghawa community, and so are wholesale distributors, so it's really easy to reach an agreement to

avoid this quarantine. (Interview with a wholesale distributor based in N'Djamena, 2021)

This reveals another factor facilitating laboratory circumvention by certain actors in the Chadian pharmaceutical market. It indeed gives a communal dimension to the circumvention. Certain members of the Zaghawa community, which he mentions, have been controlling the security apparatus (police, intelligence agency, national army, customs) for over three decades. Many of the medicine importers belong to this community, which is close to the centres of state power in Chad. These medicine importers use their relationship with members of the security apparatus close to the current regime in Chad to circumvent the directives of the regulatory authority related to the control of industrial medicines at the national laboratory. Circumvention set up by groups close to power in Chad can be analysed in the context of accumulation and power relations, as Achille Mbembe recalls in his work on power, violence, and accumulation (Mbembe 1990). The exercise of power, following Mbembe's concept, allows certain groups or individuals to gain control over resources and a free pass in various activities they undertake (Mbembe 2004, 2010). The concern for accumulation by these actors close to power allows for the observation of a form of confusion between the public and private sectors, as described in the works of Jean-François Médard (1976). In the light of these approaches developed in the work of Médard and Mbembe, if bypassing is a form of accumulation and influence, another phenomenon has its roots in the colonial context, or at least a colonial complex.

Indeed, circumvention can also be explained by postcolonial factors. In her research on the Beninese pharmaceutical market, Carine Baxerres (2011) demonstrates how francophone African countries, at the end of decolonisation, chose to import industrial medicines from France and buried, or at least did not encourage, the project of building a pharmaceutical industry like that of some anglophone countries in Africa. This also translates in the Chadian case. According to several Ministry of Health officials, medicines imported by wholesalers based in France are usually considered safe, good-quality products and are therefore not checked. This view has long been held in the French-speaking countries of Sub-Saharan Africa, but it is essential to question the trust invested in drugs imported from France. These countries often lack the human and technical resources to monitor imported medicines. Connecting drug quality to the supplier country thus becomes a justification in the discourse of local actors, even if history teaches us that the country of importation is not synonymous with the quality of an industrial pharmaceutical product. 'For a long time, the system endorsed this practice of equating quality with France, which may partly explain this circumvention' (interview with a former pharmaceutical inspector, N'Djamena, 2021). The discourse of substitution promoted by local actors represents a form of Chad's dependency on its former colonial ruler. The supply chain established by the French colonial empire thrived even after Chad's independence in 1960. According to this model, supply was exclusively conducted with French pharmaceutical industries. However, since the rise of China and India as producers of industrial medicines, the circuit has significantly diversified. As we can see, multiple reasons explain the circumventing of the laboratory facility by Chadian actors. However, circumvention is not the only form of diversion or loose appropriation of the apparatus.

THE REPURPOSING OF TECHNICAL FACILITIES

With the COVID-19 pandemic, as borders are closed, the national medicines quality control laboratory has enabled Chad to produce hydroalcoholic gels for the public and hospital supply, thanks to the contribution of the SWEDD project. (Remarks by the World Bank representative in Chad during the presentation of the first batch of hydroalcoholic gel manufactured by the laboratory financed by his institution to deal with the pandemic, N'Djamena, April 2020)

The above statement reveals another use of the technical tools initially intended for the quality control of industrial medicines in Chad. This statement also highlights the adaptability of local actors. The COVID-19 pandemic thus led to an alternative use of Chad's very first national medicines quality control laboratory. Since its establishment in 2019, the laboratory has never reached its maximum level of drug control in Chad. It is still in its infancy, and in terms of staffing (nine people), it cannot achieve the initial goal. The laboratory technicians were recruited via the SWEDD project and could only train once the technical tools had been installed on the laboratory premises. They include chemists, biologists, and public health specialists, among others. This diversity of profiles, coupled with the absence of skills to perform chemical analysis of drugs, allowed the laboratory's initial purpose to be diverted during the pandemic.

> During the pandemic, my friends and I decided to use the technical tools in the laboratory to manufacture hydroalcoholic gels for hospitals, dispensaries, and clinics – in short, for all Chadians. Thanks to the support of the Chadian government, we could manufacture over 50,000 units of hydroalcoholic gel. (Interview with head of laboratory quality control, N'Djamena, 2021)

The interview revealed the repurposing of the facility intended for drug quality control for the manufacture of hydroalcoholic gels. Given the technicians' profile, outlined above, the Ministry of Health provided additional resources for purchasing chemicals to manufacture hydroalcoholic gels. In so doing, it sought not only to demonstrate the operationality of the laboratory with a product 'made in Chad' but also to meet the needs of the population faced with a shortage of hydroalcoholic gels during the closed-border period. Against the backdrop of a global health crisis and multiple shortages, this repurposing also reflected the readaptation of a technical facility to the Chadian environment. It demonstrated the ability of local players to appropriate and modify the use of a facility designed to control industrial pharmaceuticals only. Through the national medicines quality control laboratory set up in Chad, we thus observe a bottom-up reconfiguration and readaptation of a system imposed from above. Empirical data also show the need to distinguish between a promise made by the authorities when they agreed to implement the technical facilities and the actual use of those facilities. As the wholesale distributor based in N'Djamena commented: 'It's more a matter of keeping up appearances for its financial and technical partners than of controlling medicines'. But wasn't it urgent for Chadian people to use hydroalcoholic gel?

CONCLUSION

The galloping demography of the Chadian population[2] is resulting in a growing need for medicines, which will increase the already substantial flow of pharmaceutical products and the role of transnational players in controlling supply circuits. International organisations, including primarily the World Health Organisation and the World Bank, have historically played a central role in the securitisation of medicines in the Global South. They have been investing in the (re)construction of the Chadian pharmaceutical market, exerting their influence on it through the reshaping of traditional control institutions: a pharmaceuticals inspectorate, a health police, and a regulatory authority. All these institutions have now been permeated by international organisations, through staff training, provision of material resources, or the diffusion of normative principles promoting the securitisation of Chadian pharmaceutical markets. Yet the means provided to ensure the proper functioning remain insufficient. The presence of international organisations in the Chadian pharmaceuticals market therefore appears primarily as a manifestation of the new reconfiguration of global health policies, one pillar of which is the pharmaceutical system and its securitisation.

However, this case study also shows that the Chadian government can navigate through the interests manifested by international actors and external funders. In this regard, the implementation of the lab is significant of the appropriation, circumvention, and diversion that happen in the Global South regarding international standards and funds. The Chadian case illustrates, first, how resource-poor states oscillate between signs of good will and practices of diversion; it also shows how the repurposing of the technical equipment meant to control and secure the flow of industrial pharmaceuticals brings Chad into the global medicine economy in a singular way. Retracing the genesis of the technical facility for controlling the quality of medicines and its uses shows that the technical facility has been reappropriated, repurposed, or even circumvented due to the various power dynamics between the actors involved. Adopting such a control apparatus has been far from a top-down process and has resulted in the creation of a lab whose functioning is very distinct from international standards and largely embedded in the Chadian context.

Technoscientific globalisation thus appears not only as a matter of top-down, global-local, and unidirectional exchange, but as a back-and-forth process and a series of adjustments. Despite the significant influence of international actors in disseminating standards and technical tools in the Global South, the Chadian case demonstrates that local actors do not align entirely and have a margin of action. Whether these global processes are just slowed down or entirely reinvented by local populations, however, is to be interrogated. Observing the evolution of technical devices for the quality control of medicines in Chad suggests that the country, rather than 'integrating' the global pharmaceutical markets through the one-way adoption of standardised devices and modes of control, will keep on developing its own model of drug monitoring and its own pharmaceutical rules. To take this into account, further research will have to pay increased attention to the agency of Global South players rather than documenting only the ways in which they are dominated by the most powerful actors on international markets.

ENDNOTES

1 Rapport de la politique pharmaceutique commune en zone CEMAC, acte additionnel n°07/13-CEMAC-OCEAC-CCE-SE-2.
2 18 million people according to the World Bank as of 2024: https://data.worldbank.org/indicator/SP.POP.TOTL?locations=TD

REFERENCES

Baxerres, C., 'Pourquoi un marché informel du médicament dans les pays francophones d'Afrique?', *Politique africaine*, 123 (2011): 117 https://doi.org/10.3917/polaf.123.0117.

Baxerres, C., 'Le discours sur les faux médicaments: maintenir la domination du marché pharmaceutique au temps de la libéralisation de la distribution', *Sciences Sociales et Santé*, 33 (2015): 1–23.

Beaudevin, C., and L. Pordié, 'Diversion and Globalization in Biomedical Technologies', *Medical Anthropology*, 35.1 (2016): 1–4.

De Vaujany, F.-X., 'Pour une théorie de l'appropriation des outils de gestion : Vers un dépassement de l'opposition conception-usage', *Management & Avenir*, 9 (2006): 109–26 https://doi.org/10.3917/mav.009.0109.

Desclaux, A., and M. Egrot, *Anthropologie du médicament au Sud. La pharmaceuticalisation à ses marges* (L'Harmattan, 2015).

Desclaux, A., and J.-J. Lévy, 'Présentation : Cultures et médicaments. Ancien objet ou nouveau courant en anthropologie médicale ?', *Anthropologie et Sociétés*, 27.2 (2003): 5–21 https://doi.org/10.7202/007443ar.

Fassin, D., 'La vente illicite des médicaments au Sénégal : Économies "parallèles"', *Politique africaine*, 223 (1986): 130–48.

Fassin, D., *Les enjeux politiques de la santé: Études sénégalaises, équatoriennes et françaises* (KARTHALA Editions, 2000).

Hornberger, J., 'From Drug Safety to Drug Security: A Contemporary Shift in the Policing of Health', *Medical Anthropology Quarterly*, 32.3 (2018): 365–83 https://doi.org/10.1111/maq.12432.

Mahamé, S., and C. Baxerres, 'Distribution grossiste du médicament en Afrique: fonctionnement, commerce et automédication. Regards croisés Bénin-Ghana', *Les actes des rencontres Nord/Sud de l'automédication et de ses déterminants*, 14 (2015) http://automed.hypotheses.org/Cotonou2015.

Mbembe, A. J., 'Pouvoir, violence et accumulation', *Politique africaine*, 39.1 (1990): 7–24 https://doi.org/10.3406/polaf.1990.5393.

Mbembe, A. J., 'Essai sur le politique en tant que forme de la dépense', *Cahiers d'études africaines*, 44.173–174 (2004): 173–74 https://doi.org/10.4000/etudesafricaines.4590.

Mbembe, A. J., *Sortir de la grande nuit: Essai sur l'Afrique décolonisée* (La Découverte, 2010).

Médard, J.-F., 'Le rapport de clientèle : Du phénomène social à l'analyse politique', *Revue française de science politique*, 26.1 (1976): 103–31 https://doi.org/10.3406/rfsp.1976.393655.

Morigi, V. J., and R. da S. Braga, 'La construction des savoirs sur les genres: Appropriations de la culture populaire dans les clubs de forró électronique de Fortaleza (Brésil)', *Études de communication. Langages, information, médiations*, 42 (2014): Article 42 https://doi.org/10.4000/edc.5672.

Peterson, K., *Speculative Markets: Circuits and Derivative Lives in Nigeria*. (Duke University Press, 2014)

Quet, M., 'Sécurité pharmaceutique, technologie et marché en Afrique', *Revue d'anthropologie des connaissances*, 10.2 (2016): 197–217.

Quet, M., 'Pharmaceutical Capitalism and its Logistics: Access to Hepatitis C Treatment', *Theory, Culture & Society*, 35.2 (2018): 67–89 https://doi.org/10.1177/0263276417727058.

Quet, M., *Illicit Medicines in the Global South. Public Health Access and Pharmaceutical Regulation* (Routledge, 2022).

Quet, M., and others, 'Regulation Multiple: Pharmaceutical Trajectories and Modes of Control in the ASEAN', *Science, Technology and Society*, 23.3 (2018): 485–503 https://doi.org/10.1177/0971721818762935.

Radnóti, M. *Marche forcée: oeuvres, 1930-1944.* (Phébus, 2000)

Rawicz, S. *A marche forcée : A pied, du cercle polaire à l'himalaya 1941-1942* (Phébus, 2011)

Trépied, B., 'La décolonisation sans l'indépendance ? Sortir du colonial en Nouvelle-Calédonie (1946-1975)', *Genèses*, 91.2 (2013): 7–27 https://doi.org/10.3917/gen.091.0007.

van der Geest, S., 'Les médicaments sur un marché camerounais : Reconsidération de la commodification et de la pharmaceuticalisation de la santé', *Anthropologie et Santé*, 14 (2017): 18 https://doi.org/10.4000/anthropologiesante.2450.

van der Geest, S., and S. R. Whyte, 'Popularité et scepticisme: Opinions contrastées sur les médicaments', *Anthropologie et Sociétés*, 27.2 (2003): 97–117 https://doi.org/10.7202/007448ar.

van der Geest, S., S. R. Whyte, and A. Hardon, 'The Anthropology of Pharmaceuticals: A Biographical Approach', *Annual Review of Anthropology*, 25 (1996): 153–78 https://doi.org/10.1146/annurev.anthro.25.1.153.

Whyte, S. R., 'Pharmaceuticals as Folk Medicine: Transformations in the Social Relations of Health Care in Uganda', *Culture, Medicine and Psychiatry*, 16.2 (1992): 163–186 https://doi.org/10.1007/BF00117017.

Whyte, S. R., and others, *Social Lives of Medicines* (Cambridge University Press, 2002).

CREATING ALTERNATIVE VALUES

9

MAKING VALUE OFF-PATENT: INDIA'S PHARMACEUTICAL GLOBALISATION

Yves-Marie Rault-Chodankar

'IN THE WESTERN WORLD, IF YOU MAKE SOMETHING FOR 1 DOLLAR AND you can't sell it for more than 25 dollars, it is not viable' (ARTE Reportage 2019). Talking about India's role as a primary supplier of generic medicines in the developing world, the executive director of the Serum Institute India – the largest vaccine manufacturer in the world – highlighted that the value-making practices of India's multinational pharmaceutical companies contrast sharply with their counterparts from the 'Western world'. He explained that while firms from the Global North generate income through intellectual property and monopolies, Indian firms focus on cost-effective production and wide accessibility, challenging traditional notions of value in the pharmaceutical sector.

In this chapter, I explore how value is created through the development, manufacturing, and distribution of off-patent medicines. Between 2015 and 2023, I visited many factories, offices, and warehouses across India to document the expansion of the local pharmaceutical industry. Drawing on interviews with several company directors and executives, I delve into the business models and revenue-generating strategies India's firms employ. The sample includes a wide range of enterprises: small and medium-scale businesses; biological and chemical firms; bulk drugs producers and generic formulators, manufacturing, and marketing companies; as well as international firms and India-focused businesses.

However, these firms share one common characteristic: they generate revenue from drugs originally developed in the West, which have since lost their patent.

They form an industry that contrasts sharply with the 'intellectual property-driven knowledge capitalism' that has captured researchers' attention to the detriment of the 'IP-renting countries', typically in the Global South (Kang 2021). Most studies assume that when patented drugs get stripped of their intellectual property (IP) rights, they become 'de-assetised' (Bourgeron and Geiger 2022), that is, no longer 'owned or controlled, traded, and capitalised as a revenue stream' (Birch and Muniesa 2020: 2). Hence, we rarely consider the generic industry as a critical value-making site, but rather we see it as the key to democratising access to medicines and an ally in the struggle against patents alongside activist and humanitarian actors (Biehl 2007; Lakoff 2004). From this perspective, India is the 'pharmacy of the developing world', supporting pharmaceutical affordability in the Global South against the backdrop of intellectual property laws (Guennif 2013).

Exploring alternative models of pharmaceutical value creation is thus critical in global health research (Cassier 2023). The IP-rich industry has been extensively critiqued for its inherent practice of turning lifesaving products into objects of speculation and capital accumulation (Dumit 2012; Helmreich 2008; Rajan 2006). The strategies of multinational companies from the North typically involve IP rights (Zeller 2007), which help establish 'monopoly rents' by granting rights of exclusion to the IP owner (Geiger and Gross 2021; Vezyridis and Timmons 2021). Hence, alongside research and development efforts, drugs are made profitable through the intensive management of legal and regulatory domains to create and extract value (Geiger and Finch 2016; Roy 2020). Such IP strategies are considered responsible for excessive prices and deteriorated pharmaceutical access. They are said to be at the root of the 'scarcity regime' of the Global North (Roy and King 2016). Moreover, although the expiration of patents generates new prospects for patients worldwide, it also brings about new capitalistic forms of appropriation (Hayden 2007, 2008, 2010, 2013). In fact, 'generics are compatible with both the privatisation and the pharmaceuticalisation of public health' (Hayden 2007: 488). Moreover, many interactions between IP-protected and generic companies (e.g., licensing patented drugs) blur the boundaries within the global pharmaceutical industry (Nouguez 2017; Rault-Chodankar 2022).

To contribute to these debates and illustrate how the value-making strategies of India's firms shape an alternative form of globalisation, this chapter starts by unveiling the existence of original value-making practices beyond the IP-rich industry of the Global North. I then dive into my fieldwork data to show how India's companies resort to inventive copying to generate some form of intellectual property that does not rely on patents. Next, I discuss how they diversify and adapt their manufacturing capacities to cater to small market niches, which is not profitable for other multinational firms. Then, I show how they develop distribution networks in the Global South through non-contractual relationships with suppliers, wholesalers, retailers, and doctors, contrasting with the asymmetrical relationships that IP-based firms maintain with their buyers and suppliers. Finally, I suggest that these value-making practices reflect the rise of another form of globalisation that is not based on global monopolies and large-scale concentration of power but is instead driven by local market dynamics and deeply rooted in social and economic contexts.

PHARMACEUTICAL VALUE-MAKING BEYOND THE GLOBAL NORTH

Large multinational pharmaceutical companies' capitalisation on IP rights, such as patents, is well documented (Kang 2021). The process starts with significant investment in research and development (R&D) to discover and formulate new drugs and treatments. Once a new product has been created, the company applies for a patent to secure exclusive rights to the product for a defined period, usually 20 years from the filing date. It can also prolong the legal protection of its drugs by making minor changes to the initial invention, a practice known as 'evergreening'. These exclusive rights let the company market and sell its products at premium prices and enjoy a global monopoly on its sales, earning substantial profits (Kang 2020). Companies may also license the patented drugs to other companies for additional revenue without investing in production or marketing. At the same time, they heavily invest in legal counsel and regulatory affairs to enforce their

IP monopolies. By having a well-established and exclusive distribution network, they can retain control over pricing and keep out potential competitors (they may also acquire the latter). Establishing exclusive relationships with suppliers and buyers also happens on a political level, with large companies lobbying for the establishment of international and national standards that help disqualify several competitors' pharmaceuticals as 'fake drugs', labelling them as 'counterfeit', 'substandard', or 'spurious' (Hodges and Garnett 2020).

The pharmaceutical companies' strategies do not all involve new scientific discoveries protected by the law or exclusive market presence. Outside the IP-based industry, notably in the Global South, the firms build monopolies that are more short-term, fluid, and local than those based on patents and brands, manufacturing technology, and global distribution networks. To enforce these monopolies, they do not rely on official standards but rather reappropriate them, thus giving shape to an alternative capitalist regime in which legal resources do not constitute a critical form of capital (Quet 2018). In the Global South context, contractual relationships are rarer, and business networks are embedded in various webs of social relationships. As such, they contrast with the large-scale strategies of logistics- and capital-rich multinational corporations that build on corporate, institutionalised, and contractual networks (Durand and Milberg 2020). Hence, the monopolies they create are often temporary and unstable and generate limited income for a short period. In contrast with the IP-capital-rich industry, these strategies yield a lower sale value and revenue. They are many steps removed from the emerging 'intellectual-property monopoly capitalism', led by large and sprawling technology-intensive companies (Rikap 2022a, 2022b) that dominate the hyper-specialised companies lower in the value chain (Schwartz 2022).

Companies which do not work with patents can have a more short-term focus, diversified strategies, and a tolerance for lower profits. In a way, such strategies seem to cater to the needs of patients otherwise left out by the IP-based industry. For example, Logan Williams explored an ophthalmology technology invented in the Global North that companies did not develop due to a lack of economic potential. She showed that Nepalese cataract surgeons were now using this technology as an inexpensive and convenient medical device with economic potential in South Asia (Williams 2017). While the pharmaceutical sector is not short of recent examples of value being 'shifted' rather than 'created' through patents (Reinhardt 2016), the Southern generic companies have ways of generating income by catering to the needs of a limited number of patients, something that large IP-based companies are unable or unwilling to do. Table 1 illustrates how pharmaceutical companies can operate differently across several segments of the pharmaceutical value chain.

IP-BASED PRACTICES	OFF-PATENT PRACTICES
Research & Development	
Discovering, developing, and testing new molecules, expecting technological innovation	Developing and modifying existing formulations, expecting increased drug consumption
High investment in patent and copyright enforcement	Saving costs on legal fees but spending more on marketing efforts
Building brand loyalty with first-mover advantage and based on the company's geographic origin (high symbolic capital)	Developing drug presentation with unique dosage, packaging, design, and information for niche markets
Manufacturing	
Obtaining global manufacturing authorisations: US FDA, PIC/S	Obtaining various authorisations: CDSCO, then WHO, then US FDA and PIC/S
Focus on the single-technology effectiveness of specialised production systems	Focus on the general cost and quality effectiveness of diverse and adaptable production systems
Creating value through the capacity to produce high-value pharmaceuticals	Creating value through the price of land; capacity to produce large quantities

IP-BASED PRACTICES	OFF-PATENT PRACTICES
Distribution	
Exclusive agreements with a small number of major suppliers, wholesalers, retailers, and doctors	Exclusive and non-exclusive agreements with multiple and small-sized suppliers, wholesalers, retailers, and doctors
Contractual relationships and professional network of the company	Informal relationships and critical reliance on social ties
Enjoying exclusivity by default thanks to patented products and brands	Adapting products to niche customer needs to create product singularity

TABLE 1 Comparison of value-making practices in the pharmaceutical industry

Having demonstrated the existence of original pharmaceutical value-making practices outside the IP-based industry, it would, however, be a mistake to encapsulate the Global South into a uniform model of 'copycat capitalism,' as opposed to the 'global rentier capitalism' of the Global North (Cassier 2019). The Global South countries showcase a wide diversity of situations correlated to the role of the state and the characteristics of local economic actors. In Mexico, early compliance with international intellectual property law has led expensive, foreign-made, patented drugs of the IP-based industry to dominate the market, leaving little room for the local private sector (Hayden 2008). In Nigeria, the historical merging of small-scale companies into more prominent groups to achieve economies of scale has limited the presence of low-margin drugs and the development of the local pharmaceutical industry (Peterson 2014). In Brazil, multiple drug companies, often publicly owned, have emerged by producing copies of drugs patented by multinationals from the Global North, thanks to the country's unique patent regime and the support of the state (Cassier and Correa 2009; Lakoff 2004). In contrast, India's pharmaceutical industry has developed rapidly thanks to its private sector, which has benefitted from the protectionist measures of the state and lack of adherence to the agreement on

Trade-Related Aspects of Intellectual Property Rights (TRIPS) before 2005. This unique political-economic assemblage now shapes the idiosyncratic value-making practices around off-patent drugs.

'ALPHONSO MANGO INSTEAD OF A MANGO': DEVELOPING ALTERNATIVE INTELLECTUAL PROPERTIES

India's pharmaceutical industry has over 10,000 companies of various sizes, including approximately 485 companies with export licenses[1], with only a few listed on the stock exchange. They generally cannot afford to invest in drug discovery and development activities as large multinational firms. While R&D-related expenses amount to 8–15% of costs for European and North American pharmaceutical companies, they represent a tiny share of the annual turnover of India's companies (Joseph 2016). These limited R&D investments are made by the most prominent companies, such as Dr Reddy's and Biocon, and generally go towards developing generic versions of off-patent pharmaceuticals. One former executive of the large Indian companies Dubar, Intas, and Alchem, who now manages a small pharmaceutical distribution company based in Mumbai, explained that:

> In India, there is significantly less spending on R&D. The big companies only invest 2–3% of their turnover. Among the 20,000 companies in India, only ten companies are spending on R&D: Dr Reddy, Dabur, Glenmark, Sun Pharma, CIPLA [...] Why would they spend so much money when there are so many molecules available in the market?

In this context, 'molecules' refers to pharmaceutical compounds, particularly active pharmaceutical ingredients (APIs), that are already available for production. Rather than investing heavily in discovering new drugs, many Indian pharmaceutical companies focus on developing cost-effective generic alternatives. Even though some of India's leading companies present themselves as innovation-driven and make more significant investments in R&D activities, their model primarily relies on developing and producing cost-effective generic alternatives. They rely, in a way, on alternative forms of intellectual property. Biocon is one of these companies, which heavily invested in the development of original biologic drugs (especially between 2002 and 2011) such as Nimotizumab and Itolizumab, two novel monoclonal antibodies, before going back to developing equivalents of existing cancer drugs in partnership with a US multinational firm. A former R&D executive at Biocon, who partnered with a Cuban research company to develop these products, explained that investing in R&D for novel drugs was ultimately part of the company's strategy to build expertise in the development and manufacturing of biosimilar drugs:

> The business model of Biocon was overall generics, so if you look at it, we developed small molecules. From there we moved to biosimilar insulins, and then we needed to get learnings on monoclonal antibody therapeutics. So, we got into this fantastic collaboration with CIM [Cuban company], and we learned that, and we used all the learning to run the biosimilar business. I strongly believed that the focus on novel drugs was there, but the return on investment was not enough.

In the past decade, most of Biocon's profit has come from the sale of widely consumed biosimilars such as insulin glargine, pegfilgrastim, and Trastuzumab, which the company was able to develop as pharmaceutical equivalents and produce at a much lower cost. With these strategies, the largest Indian generic companies can hope to achieve a 'first-to-file' monopoly, that is, the right – granted to the first mover in certain countries like the United States – to sell exclusively at any price and for a limited period of time a product that is bioequivalent to the princeps drug. For the rest of the Indian companies, R&D activities rarely involve modifying an existing molecule; instead, they focus on changing how it is presented to the buyers and prescribers, such as pharmacists, doctors, and patients. These products constitute the firms' 'portfolio', proudly displayed in their offices (see Figure 9.1).

Consequently, many of India's drug development activities in the private sector are geared towards turning generic drugs into unique products by changing the dosage, mode of administration, and branding of the products, often

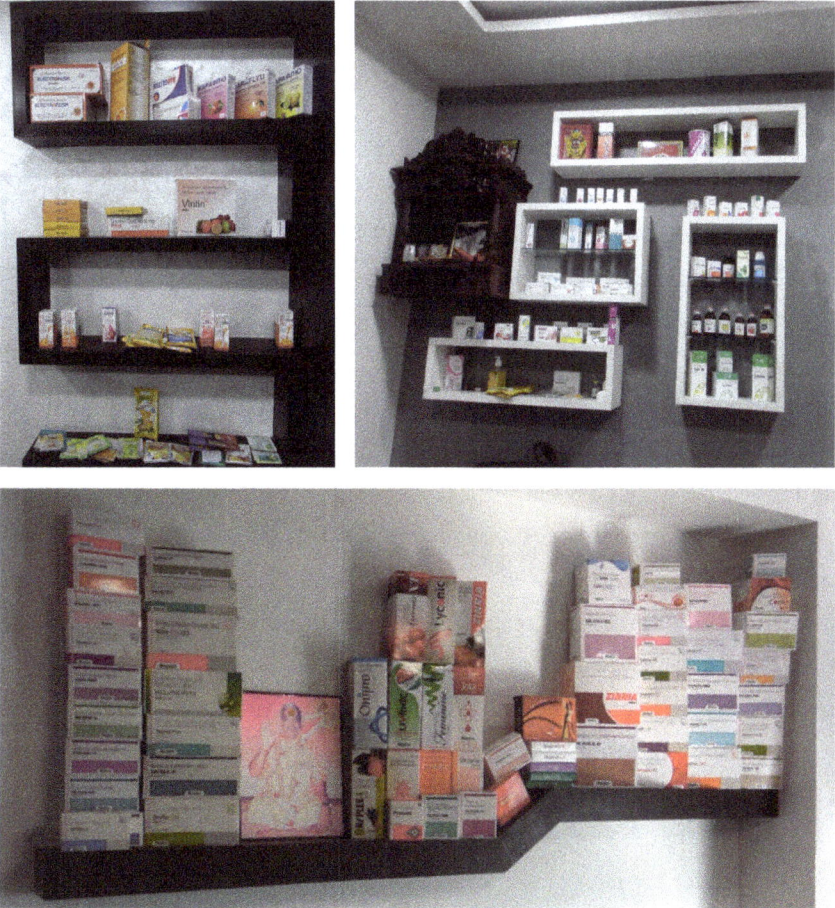

FIG. 9.1 Product display in three corporate offices (Yves-Marie Rault-Chodankar)

known as 'me-too drugs'. One company's strategy was to improve the dissolution profile of a formulation. Its CEO explained that rather than merely propose a cheaper product, she was developing a more affordable and better product, or in her words, an 'Alphonso mango instead of a mango' (a tastier variety of mango). Thus, one active area of India's incremental innovation has been novel drug delivery systems (NDDS). Successful examples include an Indian-improved version of the antibiotic product Ciprofloxacin, which challenged the dominant position of the multinational firm Bayer in early 2000. One R&D executive, who recalled his participation in the development of a treatment only available as an injection into a topical gel, nevertheless confirmed the scarcity of radical pharmaceutical innovation in India.

> There is some innovation in NDDS, but it is not a big thing. Apart from Zydus, no Indian pharmaceutical company has developed a molecule in ten years. The market is filled with generic drugs because the patent law in India was only applied after 2005. The companies can only come with new brands from existing generic drugs.

NDDS are generally tricky to protect through the courts, unlike product patents that retain their value because large firms can build robust legal defences (Geiger and Finch 2016). As a result, NDDS can only secure monopolies for a short period until other companies copy the product. Drug copying usually happens once the drugs are adopted and known by enough patients for their copy to be profitable. Another way to create differentiation through copying is to change the dosage of generic medicines. The director of a medium-sized company was planning to commercialise a drug for pre-menopausal women (Vagifam) at a lower dosage to create a short-term monopoly:

> Somebody gave us this product for development. [...] This product, Vagifam, is used for pre-menopausal women. It's a single-use formulation which women need to use in their vaginas to reduce the symptoms of menopause. Now, the usual dose was 25 mg, but for no reason, no one decided to introduce this only in 10 mg. [...] They each will cost 4–5 euros. So, the cost of the drug, in this case, is very minuscule.

These 'supergenerics', named as such because of their use of 'sameness and similarity as generative forms of distinction and value' (Hayden 2013: 601), are often prescribed, sold, and purchased under brand names. The presentation and branding of drugs is thus a significant area in which India's generic companies have innovated in recent years, capitalising on colours, shapes, designs, and packaging materials, as well as including detailed information about the product on the label, to foster a sense of reliability. In parallel, other companies have started reemphasising the generic aspect of their products, for instance, by developing minimalist packaging or emphasising the lower prices of 'pure generics' to appeal to specific population categories, even if their prices are not necessarily lower than the competition.

This drive to create unique generic products has led to the proliferation of multiple pharmaceutical products across India's markets. In June 2023, the Indian pharmaceutical marketplace 1mg.com offered over 12,300 paracetamol-based products in various forms involving various dosages and combinations, commercialised under more than 1,100 different brands. However, these brands survive not by relying on copyright enforcement, exclusive agreements, and monopoly-building but by catering to diverse market needs and patient niches. Most small-scale Indian companies choose not to register their brand names to save on some administrative and legal consulting fees. However, while intellectual property is virtually absent from their strategies as suppliers of generic drugs, the Indian companies have to contend with other regulations to expand their production capacity, especially manufacturing standards. The following section documents how Indian companies create value through manufacturing strategies that differ significantly from those of large foreign multinationals.

AMLA CANDY, COUGH SYRUPS, AND NICHE MARKETS: THE VALUE OF FLEXIBLE MANUFACTURING

India now hosts over 2,000 pharmaceutical factories certified by the World Health Organisation[2]. It has the highest number of US FDA-approved drug-manufacturing plants outside the USA[3]. Since the 1970s, investors from various sectors have seen India's pharmaceutical manufacturing industry as a safe and dynamic economic space and an opportunity to diversify their portfolio. So much so that the overall manufacturing capacity of India's pharmaceutical companies has not been fully utilised, and several plants have been shut down in the last 20 years as they failed to adopt national and international manufacturing practices while remaining profitable (e.g., India's Schedule M; see Iyer 2008).

The surviving manufacturing plants could upgrade their production processes to comply with international manufacturing rules. One factory from

Himmatnagar (Gujarat), set up in 2014, which produced specialised injections, was, however, running at only 10% of its capacity due to a lack of buyer orders. The facility would have been worth little if the value was based on its current usage. However, the CEO explained that the company's sales value had continuously increased since it obtained its first manufacturing license and anticipated that its upcoming PIC/S certification (standards of the European Free Trade Association) would increase its sales value even more. He therefore had to keep a costly workforce on the payroll, including compliance and quality officers, even though production was insufficient to make the company's operations profitable. Thus, to hold onto certifications and maintain the company's value, the managers paradoxically had to create an unprofitable manufacturing system. When I met him in 2017, the CEO explained that the plant should now be worth 45 crore rupees (5.7 million USD) when they had invested 30 crore rupees (3.8 million USD), even though the factory was not economically viable. The CEO explained that this was a long-term strategy supported by local investors looking to diversify their assets:

> For this kind of company, you should not be looking for short-term profits. At first, we had difficulty finding investors, but we found one big investor with the Vohra Samaj. The Harsolia family is one of the big families that invested, and one son is now in charge of the financial department. They are wealthy because they have shares in the automobile sector, but they wanted to develop a vertical in pharmaceuticals.

This drive to diversify investments is a visible trend within pharmaceutical organisations. For many companies, a key strategy is to comply with specific manufacturing standards (chiefly, Good Manufacturing Practices – GMPs) rather than focus on only the most demanding ones, like the IP-rich pharmaceutical industry. Indian companies often follow an inverse process, first obtaining less stringent certifications from local agencies, then meeting World Health Organisation (WHO) standards to access donor-based markets in the Global South, and eventually complying with the most stringent standards set by the USA or the EU. The GMPs are constantly changing, and as they do, the current GMPs (cGMPs) also affect the value of companies, for compliance with these practices determines which markets a company can access and the price at which it can sell pharmaceuticals. The inflation of norms has led to a race to obtain and hold onto certifications, reflected in the growing number of regulatory consultants specialised in compliance with the production standards of the various health authorities.

Manufacturing is not fundamentally driving valuation in India, especially compared to the United States (Rajan 2006: 127). The Indian firms' strategy instead consists of validating their manufacturing capability, that is, their ability to produce and sell multiple products in competition with multinational firms (Hughes 2013). The strategies of Indian companies generally focus on improving manufacturing capacities rather than developing ways to produce innovative pharmaceuticals. For instance, a manufacturing company from Ambernath (Maharashtra) could access the European market as a supplier of API through a

FIG. 9.2 *Left*, a cGMP-compliant factory, inaugurated in 2019; *right*, a factory being constructed, Pune, February 2023 (Yves-Marie Rault-Chodankar)

so-called written confidential with an Austrian company. The European company vouched that the plant complied with the PIC/S certification requirements for importing Indian products. However, the head of business development explained that they were still applying for the PIC/S certification to diversify their customer base but also to find another source of valuation for his company:

> We cannot produce for the domestic market because there is too much cost compression, and eventually, our products will get a bad purity profile. This is why we mostly work in exports with partners from Europe or the US. [...] For now, we don't have the big certifications, but I can tell you the factory is already compliant. But we are looking to get the PIC/S certification even if we don't really need it because it will give us more credit [credibility].

Another manufacturing plant in Sanand (Gujarat) produced amoxicillin and potassium clavulanate tablets for two companies, one Indian and the other British. The CEO assessed the value of the company's plant against the income he could generate by producing the same product for different companies since it did not require further investment. In contrast with the hyperspecialised production lines for patent technologies, the Indian plants can integrate the production of diverse pharmaceuticals to maximise their capacity and revenue. Some Indian companies notably developed small-scale systems to produce monoclonal antibodies to cater to the limited domestic demand for these biological drugs. The most iconic example is probably the case of a small-scale plant from Ahmedabad, which was simultaneously manufacturing amla (Indian gooseberry) candy and cough syrups to make full use of its production capacity. The following section goes one step further in the pharmaceutical value chain to understand how Indian companies create value by distributing drugs to the Global South.

THE 'MINUSCULE NETWORKS': DISTRIBUTING TO THE REST OF THE WORLD

A significant portion of India's pharmaceutical revenue is derived from the Global South, or to use the industry's terminology, the 'Rest of the World'. In the year 2020, sales on the domestic market accounted for 50% of the industry's revenue, while sales in other low- and middle-income countries (LMICs) contributed 25% of its overall earnings. Large foreign companies often handle the distribution of Indian-made products to high-income markets. Complying with existing regulations, convincing medical practitioners, and negotiating with insurance companies are complex and costly activities that Indian firms often cannot manage alone. Instead, they focus on distributing their products to 'semi-regulated' markets in the Global South. The Indian brands must develop unique strategies, as the geographical origin and associated perceptions of quality have a considerable impact on the consumption of generics, and brands from India usually have a negative reputation (Baxerres, Kpatchavi, et al. 2021).

Access to the markets of the Rest of the World is facilitated by many intermediaries, which Indian companies leverage to distribute their products. India's business networks are significantly based on family, community, regional, and socio-professional ties, a dynamic that profoundly permeates how South–South networks are shaped (Rault-Chodankar 2020). The companies rely extensively on their social ties to secure, protect, and control networks in a Global South context, where value chains are governed by multiple 'middlemen' or 'traders', in contrast with North–South networks characterised by power asymmetries (Horner and Murphy 2017). Networking is paramount in India's pharmaceutical industry, with companies relying extensively on social ties to secure and control market access. For instance, the CEO of an Ahmedabad-based company used a family connection in Ghana to find a distributor for his antidiabetic products, which were manufactured in his uncle's factory in Baddi (Himachal Pradesh). Additionally, to forge new connections, visiting cards (see Figure 9.3) are frequently exchanged during industry events, demonstrating another way Indian firms make contact and build crucial networks.

Contacts in international markets are particularly crucial to learning about local manufacturing and commercial standards. One manager found a Danish consultant from India, certified in clinical studies, to help his company register a generic version of a drug treating menopausal symptoms on the EU market. Yet other companies generated revenue through their networks by obtaining exclusive information about missing products in various markets. One company

FIG. 9.3 Some visiting cards collected during fieldwork (Yves-Marie Rault-Chodankar)

identified a treatment for polycystic ovary syndrome and irregular menstrual cycles in women that was still sold only in Europe and the USA. These examples illustrate the strength of small ties in India's pharmaceutical industry, as confirmed by this company director based in Ahmedabad, who recalled how helpful his network has been in the early days:

> I had a friend, a chartered accountant, who gave me advice in accountancy. He gave me very simple books on finance. They were caricature books, comic books on finance. I enjoyed, and I understood how balance is made. And it is my friend from college days who taught me how to make project reports and how to borrow money. If you look at my networks, these are my minuscule networks which have built me.

Certain companies with more limited 'social capital' struggled to access supply chains and distribution networks. For instance, a newly built factory in Gujarat could not find buyers for its specialised injections. The CEO, a Muslim community member whose local board owned the company, argued that they were excluded from Hindu networks because they were Muslim. Consequently, they could only manufacture for large Muslim-owned companies like Cipla. He explained that most of the social networks of his Muslim caste (Bohra) were still in the automotive parts manufacturing industry. Another manufacturer of low-cost pharmaceuticals, such as cough syrups, based in Surendranagar (Gujarat), stated that being in a city that was not a business hub was a significant disadvantage. The manager had to travel weekly to Ahmedabad, two hours away by car, to meet potential buyers. With a limited command of English and knowledge of information technology, he explained that he struggled to find customers.

> I travel weekly to Ahmedabad to meet customers because they do not come to Surendranagar. Other firms from Ahmedabad, they just talk to their neighbours and find connections, so it's hard for us. They find marketing firms and sell their products. For us, we tried to improve our marketing, but the internet is not so good in Surendranagar. We have paid GoDaddy to create a website, but it does not work.

The Indian companies also leverage their social networks to build exclusive relationships with distributors. In India, whether they work in private clinics or public hospitals, doctors generally prescribe not the generic but the branded version of a drug, despite the medical associations' recommendations and unsuccessful legislative attempts to encourage the prescription of generic medicines. Promoting brands is not so much a matter of convincing customers of the quality of a given product but instead of ensuring that the hospitals, pharmacists, and doctors will not even consider another brand. Consequently, many companies' activities involve promoting products to those influencing their sale or prescription. Due to physicians' prescription practices, pharmacies and hospitals must store specific brands. At the same time, pharmacists can also bypass prescriptions and sell prescription drugs without an official document from a doctor (Dahdah et al. 2018). If a doctor prescribes a generic brand name, the pharmacists will choose the product from a series of generic brands with the same chemical formula. Building solid and exclusive ties with medical practitioners is thus a significant way of securing industry and regional monopolies. Pharmacists are also the leading partners in selling over-the-counter drugs when a doctor's prescription is not required. Furthermore, having a local distribution network provides companies with valuable data on local customer preferences, buying patterns, and market trends, which they can use to make informed business decisions and develop new strategies. The companies can also use this information to tailor their marketing efforts to local customer needs and increase customer loyalty.

Although some medical-scientific arguments are commonly used to highlight the unique benefits of specific dosages, India's companies mainly draw on preexisting scientific data to promote cheaper branded variations of a molecule. These practices differ from the 'scientific marketing' activities described in markets of the Global North, designed to generate new needs and to convince the medical community, including health agencies and officials at different levels (Gaudillière 2015). For example, Greffion and Breda (2015) argue that French medical representatives (MRs) primarily promote the prescription of the costliest and most recent drugs. In India, the far wider variety of low-cost brands, the limited number of people covered by health insurance, and the absence of social security suggest a different role for MRs, geared towards convincing not medical authorities and private and public insurance to accept new molecules, but rather the patients, doctors, and retailers to adopt different versions of existing products.

A country like Ghana is also host to these lively promotional practices, where 2,175 private pharmacies are supplied by 576 private wholesalers, a situation which contrasts drastically with that witnessed in Benin, where a few wholesaler distributors have a strong monopoly (Baxerres, Codjo, et al. 2021). In India, pharmaceutical companies extensively use their connections with medical professionals to make them prescribe their products. For example, the CEO of a company from Ahmedabad, a former medical representative, was promoting formulations for men with erectile dysfunction (generic Sildenafil). To do so, he developed exclusive agreements with doctors and local wholesalers to prevent them from proposing other generic alternatives. In India, former medical representatives extensively use their professional networks to create pharmaceutical enterprises. The ongoing war between marketing firms to poach one another's most experienced MRs by offering them better income and higher positions is evidence of the importance of local connections in the sector.

In this context, business consultants play a crucial role. Former executives of large pharmaceutical companies often rent out their networks as a fee-based service. The director of a pharmaceutical consultancy based in Mumbai was posted in Africa as a regional manager for an extended period and built a 20-year career in the sector. He explained how his extensive transnational network could help him find buyers and a range of helpful information that can generate economic revenue:

> Over the years, whatever network I have developed can be divided basically into three: one is my core network of core people that I know are looking for a product. They are all connected, and these people are also people who would respond to me in case. So, these are the people you have to talk to. So, believe me, if you have a contact list of 500, this core list is just 10–20 maximum, no more than that. And then there is the other category they will tell you maybe they are interested, maybe they will come back to you, maybe they will not, but they are the potential clients. And number 3 is you know some people you just have to pass on the information, and maybe if they are interested, they will come back to you.

Indian business networks are, however, not as exclusive as in the IP-based industry. For example, one company that had set up a European branch also marketed the products of other companies based in India. Local networks thus also work as connectors by facilitating interaction through shared culture, language, or community ties. By the same token, they can serve as exclusion tools to keep out competitors. Unlike in the patent-protected pharmaceutical industry, access and exclusion coexist as strategies in India's pharmaceutical networks. Extensive fluidity, the blurred boundary between the interpersonal and the corporate, and the non-contractual nature of inter-company relationships are distinct specificities of the pharmaceutical industries of the Global

South. India's firms bring business practices that shape unique technoscientific assemblages, which deeply contrast with the intellectual-property monopoly capitalism led by IP-based companies.

CONCLUSION: AN ORDINARY GLOBALISATION

This chapter examines the value-making practices of the pharmaceutical industry beyond the large and innovative multinational companies of the Global North. India's pharmaceutical business strategies illustrate how firms can access niche markets with off-patent formulations, develop cost-efficient and adaptable production systems, and form diversified distribution networks through exclusive and non-exclusive agreements with local suppliers, wholesalers, retailers, and doctors.

So far, pharmaceutical capitalism has primarily been studied from/in the Global North, looking at large multinational companies and 'high-end' processes of transnational interconnection. By examining the smaller, more localised, less radically innovative pharmaceutical companies in the Global South, this chapter answers the call to integrate actors and objects that are 'non-hegemonic' and almost 'ordinary' but still play an instrumental role in global economic change, as there are many of them and they are highly active (Mathews et al. 2012).

The dynamism of India's pharmaceutical companies illustrates the liveliness of 'low-end' globalisation processes, that is, the 'transnational flow of people and goods involving relatively small amounts of capital and informal, sometimes semi-legal or illegal transactions, commonly associated with the "developing world"' (Mathews 2011: 19–20). This concept has been instrumental in highlighting how certain social groups, typically considered marginal, actively participate in globalisation processes, even though they do not belong to 'worlding cities' or large multinational companies (Choplin and Pliez 2015).

Regarding relations of domination, it is crucial to recognise that several forms of capitalism can grow in parallel and interact without strong relations of domination. Indian firms are dominant in many ways. For instance, they play a crucial role in the global supply chain for generic medicines, often providing the bulk of affordable medications for low- and middle-income countries. Their dominance is characterised not by monopolistic control but rather by their extensive reach, adaptability, and ability to innovate within the constraints of existing technologies and markets. Indian pharmaceutical companies have mastered the art of reverse engineering and the development of cost-effective manufacturing processes, which has allowed them to thrive in both local and international markets, especially within the Global South.

These original value-making practices shape an alternative form of globalisation. Indian firms appropriate and adapt existing drugs, finding new ways to create value in local and international markets, particularly in the Global South. By leveraging off-patent pharmaceuticals, they carve out niches that larger multinational corporations often overlook. This approach not only democratises access to essential medicines but also underscores the importance of alternative, non-monopolistic forms of globalisation that significantly contribute to economic and social development.

More scholarly attention must thus be paid to non-monopolistic strategies to understand the global diversity of technoscientific capitalism. Objects without intellectual property value, such as generic pharmaceuticals, also have an intense economic afterlife, thus suggesting that globalisation studies should consider old technologies (e.g., off-patent pharmaceuticals) as driving modern capitalist history just as much as shiny, IP-protected innovations (see Edgerton 2011).

In essence, the chapter underscores the need to expand our understanding of global economic dynamics to include these 'ordinary' actors and objects who, despite their lack of hegemonic power, play a vital role in shaping the contours of global capitalism. Ordinary strategies and practices also offer valuable insights into how economic value can be generated and sustained outside the paradigms of high-end innovation and monopolistic practices. They enrich our understanding of the complex and multifaceted nature of technoscientific globalisation.

ENDNOTES

1 *List of members.* (2024). pharmexcil.com. https://pharmexcil.com/members. Accessed 1st February 2024
2 *List of WHO GMP Manufacturing units* (Central Drugs Standard and Control Organisation 2019). https://cdsco.gov.in/listwhogmp. Accessed 18th March 2025

3 *Generic Drug Facilities, Sites and Organization Lists.* (Food and Drug Administration 2019) https://www.fda.gov/industry/generic-drug-user-fee-amendments/generic-drug-facilities-sites-and-organization-lists. Accessed 18th March 2025.

REFERENCES

ARTE Reportage, *Inde: la guerre des vaccins* (ARTE, 2019) https://www.arte.tv/fr/videos/071609-000-A/inde-la-guerre-des-vaccins/.

Baxerres, C., A. Codjo, and D. K. A. Kpatchavi, 'Distribution and Access to Medicines', in C. Baxerres, and M. Cassier, eds., *Understanding Drugs Markets* (Routledge, 2021), pp. 72.

Baxerres, C., A. C. Kpatchavi, and D. K. Arhinful, 'When Subjective Quality Shapes the Whole Economy of Pharmaceutical Distribution and Production', in C. Baxerres, and M. Cassier, eds., *Understanding Drugs Markets* (Routledge, 2021), pp. 249–73.

Biehl, J., 'Pharmaceuticalization: AIDS Treatment and Global Health Politics', *Anthropological Quarterly*, 80 (2007): 1083–1126.

Birch, K., and F. Muniesa, eds., *Assetization: Turning Things into Assets in Technoscientific Capitalism* (MIT Press, 2020), pp. 1–41 https://direct.mit.edu/books/book/4848/AssetizationTurning-Things-into-Assets-in.

Bourgeron, T., and S. Geiger, '(De-)Assetizing Pharmaceutical Patents: Patent Contestations Behind a Blockbuster Drug', *Economy and Society*, 51 (2022): 23–45.

Cassier, M., 'La Fin du Partage? Les Capitalismes de la Copie Face au Capitalisme de la Rente Globale: Une Nouvelle Géographie des Industries de Santé', *Mouvements* (2019): 107–19.

Cassier, M., *Il Y a Des Alternatives: Une Autre Histoire Des Médicaments (XIXe-XXIe Siècle)* (Seuil, 2023).

Cassier, M., and M. Correa, 'Éloge de la Copie: Le Reverse Engineering des Antirétroviraux Contre le VIH/Sida Dans les Laboratoires Pharmaceutiques Brésiliens', *Sciences Sociales et Santé*, 27 (2009): 77–103.

Choplin, A., and O. Pliez, 'The Inconspicuous Spaces of Globalization', *Articulo – Journal of Urban Research*, 12 (2015): Article 12 https://doi.org/10.4000/articulo.2905.

Dahdah, M.A., A. Kumar, and M. Quet, 'Empty Stocks and Loose Paper: Governing Access to Medicines Through Informality in Northern India', *International Sociology*, 33 (2018): 778–95.

Dumit, J., 'Prescription Maximization and the Accumulation of Surplus Health in the Pharmaceutical Industry', in K. S. Rajan, ed., *Lively Capital* (Duke University Press, 2012), pp. 45–92.

Durand, C., and W. Milberg, 'Intellectual Monopoly in Global Value Chains', *Review of International Political Economy*, 27 (2020): 404–29.

Edgerton, D., *The Shock of the Old: Technology and Global History since 1900* (1st edn; Oxford University Press, 2011).

Gaudillière, J.-P., *The Development of Scientific Marketing in the Twentieth Century: Research for Sales in the Pharmaceutical Industry* (Routledge, 2015).

Geiger, S., and J. Finch, 'Promissories and Pharmaceutical Patents: Agencing Markets Through Public Narratives', *Consumption Markets & Culture*, 19 (2016): 71–91.

Geiger, S., and N. Gross, 'A Tidal Wave of Inevitable Data? Assetization in the Consumer Genomics Testing Industry', *Business & Society*, 60 (2021): 614–49.

Greffion, J., and T. Breda, 'Façonner la Prescription, Influencer les Médecins', *Revue de la Régulation. Capitalisme, Institutions, Pouvoirs*, 17 (2015) https://doi.org/10.4000/regulation.11272.

Guennif, S., 'L'Économie Politique du Brevet au Sud: Dimensions Industrielle et Sanitaire', *Mondes en Développement*, 160 (2013): 85–98.

Hayden, C., 'A Generic Solution? Pharmaceuticals and the Politics of the Similar in Mexico', *Current Anthropology*, 48 (2007): 475–95.

Hayden, C., 'No Patent, No Generic: Pharmaceutical Access and the Politics of the Copy', in M. Biagioli, P. Jaszi, and M. Woodmansee, eds., *Contexts of Invention* (University of Chicago Press, 2008), pp. 62–90.

Hayden, C., 'The Proper Copy', *Journal of Cultural Economy*, 3 (2010): 85–102.

Hayden, C., 'Distinctively Similar: A Generic Problem', *UCDL Review*, 47 (2013): 601.

Helmreich, S., 'Species of Biocapital', *Science as Culture*, 17 (2008): 463–78.

Hodges, S., and E. Garnett, 'The Ghost in the Data: Evidence Gaps and the Problem of Fake Drugs in Global Health Research', *Global Public Health* (2020): 1–16.

Horner, R., and J. T. Murphy, 'South–North and South–South Production Networks: Diverging Socio-Spatial Practices of Indian Pharmaceutical Firms', *Global Networks* (2017): 1–26.

Hughes, S. S., *Genentech: The Beginnings of Biotech* (University of Chicago Press, 2013).

Iyer, P. K., *Structure and Performance of Small and Medium Scale Pharmaceutical Firms*, SSRN Scholarly Paper, 15 March 2008 https://papers.ssrn.com/abstract=1752125.

Joseph, R. K., *Pharmaceutical Industry and Public Policy in Post-Reform India* (1st South Asia edn; Routledge, 2016).

Kang, H. Y., 'Patents as Assets: Intellectual Property Rights as Market Subjects and Objects', in K. Birch and F. Muniesa, eds., *Assetization: Turning Things into Assets in Technoscientific Capitalism* (MIT Press, 2020), p. 28.

Kang, H. Y., 'Patent Capital in the Covid-19 Pandemic: Critical Intellectual Property Law', *Critical Legal Thinking*, 9 February 2021 https://criticallegalthinking.com/2021/02/09/patent-capital-in-the-covid-19-pandemic-critical-intellectual-property-law/.

Lakoff, A., 'The Anxieties of Globalization: Antidepressant Sales and Economic Crisis in Argentina', *Social Studies of Science*, 34 (2004): 247–69.

Mathews, G., *Ghetto at the Center of the World: Chungking Mansions, Hong Kong* (University of Chicago Press, 2011).

Mathews, G., G. L. Ribeiro, and C. A. Vega, *Globalization from Below: The World's Other Economy* (Routledge, 2012).

Nouguez, É., *Des Médicaments à Tout Prix: Sociologie Des Génériques En France* (Presses de Sciences Po, 2017).

Peterson, K., *Speculative Markets: Drug Circuits and Derivative Life in Nigeria* (Duke University Press, 2014).

Quet, M., 'Diversions de Flux et Contestations du Régime', in M. Quet, ed., *Impostures Pharmaceutiques: Médicaments Illicites et Luttes Pour l'accès à La Santé* (La Découverte, 2018), pp. 187–212.

Rajan, K. S., ed., *Biocapital: The Constitution of Postgenomic Life* (Duke University Press Books, 2006).

Rault-Chodankar, Y.-M., 'Chapitre 6. Des Communautés de Ressources Économiques', in *Les Petites Entreprises Pharmaceutiques Indiennes, Agents d'une Globalization Alternative* (Université de Paris, 2020), pp. 189–232 https://hal.archives-ouvertes.fr/tel-02484731/.

Rault-Chodankar, Y.-M., 'Domestiquer la Norme Mondiale: Brevet Pharmaceutique, Bonnes Pratiques de Fabrication et Contrôle du Prix des Médicaments en Inde', *L'Espace Politique*, 45 (2022) https://doi.org/10.4000/espacepolitique.10628.

Reinhardt, U. E., 'Value Creation and Value Shifting in Health Care', *Health Affairs Forefront*, 1 June 2016 https://www.healthaffairs.org/do/10.1377/forefront.20160601.055099/full/.

Rikap, C., 'Amazon: A Story of Accumulation Through Intellectual Rentiership and Predation', *Competition & Change*, 26 (2022a): 436–66.

Rikap, C., 'From Global Value Chains to Corporate Production and Innovation Systems: Exploring the Rise of Intellectual Monopoly Capitalism', *Area Development and Policy*, 7 (2022b): 147–61.

Roy, V., 'A Crisis for Cures? Tracing Assetization and Value in Biomedical Innovation', in K. Birch and F. Muniesa, eds., *Assetization: Turning Things into Assets in Technoscientific Capitalism* (MIT Press, 2020), pp. 97–124 https://direct.mit.edu/books/book/4848/AssetizationTurning-Things-into-Assets-in.

Roy, V., and L. King, 'Betting on Hepatitis C: How Financial Speculation in Drug Development Influences Access to Medicines', *BMJ*, i3718 (2016).

Schwartz, H. M., 'Intellectual Property, Technorents and the Labour Share of Production', *Competition & Change*, 26 (2022): 415–35.

Vezyridis, P., and S. Timmons, 'E-Infrastructures and the Divergent Assetization of Public Health Data: Expectations, Uncertainties, and Asymmetries', *Social Studies of Science*, 51 (2021): 606–27.

Williams, L. D., 'Getting Undone Technology Done: Global Techno-Assemblage and the Value Chain of Invention', *Science, Technology and Society*, 22 (2017): 38–58.

Zeller, C., 'From the Gene to the Globe: Extracting Rents Based on Intellectual Property Monopolies', *Review of International Political Economy*, 15 (2007): 86–115.

10

THE DIVISION OF BIOMETRIC LABOUR: RELATIONS OF PRODUCTION IN AFRICAN VOTER-IDENTIFICATION TECHNOLOGIES

Cecilia Passanti

SINCE THE TURN OF THE CENTURY, DIGITAL TECHNOLOGIES HAVE PLAYED A growing role in public administration around the world. This is particularly visible in Africa where, in response to allegedly dysfunctional public services and inefficient identification frameworks, digital systems are being mobilised in many areas of state-citizen interactions (Breckenridge and Szreter 2012). Biometrics, understood as 'identifying people with machines' (Breckenridge 2014), is one of the most vivid illustrations of this trend. Biometric identification based on computer systems was initially used to match fingerprints taken at a crime scene with criminals already known to the police system (Cole 2009). Under the impetus of the incumbent technology manufacturers and in collaboration with increasingly computer-savvy governments, these systems have since been applied to a wide range of sectors, including border control, healthcare, and government subsidies management, as well as the production of certified machine-readable documents. In addition to this, the biometric industry provides computerised voter-list management, voter verification, and other systems to facilitate and expedite the voting process, such as voting machines and results transmission systems (RTS). Biometrics in voting is designed to make voter identification more efficient by capturing voters' biological data, namely their fingerprints and facial images. These data feed a digital list, stored in a database, which a generic biometric software programme called Automated Fingerprint Information System (AFIS) uses to compare the data in order to detect and eliminate the records of any citizens registered twice. Through this software processing, biometrics is supposed to produce more reliable lists of citizens and thus more trust in public institutions.

Although biometrics are used worldwide for elections – or for democracy, as vendors and experts often put it – they are mostly developed in and for Global South contexts, particularly Africa. Whereas many activists, journalists, and researchers see the entry of biometrics into the field of elections as an attack on the ethics of democracy, many African countries – 36 out of 56 (Debos and Desgranges 2023) – have adopted and engaged in a long-term relationship with this technology precisely to strengthen democracy. Biometrics' reciprocal love of African governments makes these systems relevant artefacts for the study of the North-South relations embedded in global technological products, and for understanding how they fuel their spread on a global scale.

Biometric identification systems are generic artefacts that move from one context to the next, across organisations (e.g., from forensic to voting) and geographic areas (e.g., from country to country) without losing their nature. They are tremendously adaptable, and their qualities can be reshaped according to local needs. While this adaptive capacity may point to the relevance of studying how identification systems adapt to local contexts (localisation and adaptation work), this chapter takes a different approach, focusing on the production (Pollock et al. 2007) of both generic biometrics and more specific electoral biometrics.

Building on debates about globalisation and postcolonial STS, I argue that the North-South relationship is inscribed in the history of the invention of electoral biometrics and reproduced in contemporary relations of production, namely the division of biometric labour. This concept refers to the structured organisation of work to produce electoral biometric artefacts, which involves the daily work of institutions and individuals at both national and transnational levels.

The division of biometric labour functions as a travelling model (Behrends et al. 2014), based on work to produce genericity (Pollock et al. 2007), that supports biometrics' ability to adapt to disparate contexts. The literature has argued that the genericness and globalisation of science and technology are achieved through work to detach them from their original social relations of production (Akrich et al. 2011). In contrast, I argue that globalisation arises from the reiteration and objectification of their original social relations of production.

The questions guiding this research stem from a debate about the role of Africa, colonial, and postcolonial situations in the emergence of biometric sciences and technologies. Breckenridge framed the African origins of biometric sciences and the resulting statistical-mathematical model of governance (Breckenridge 2014, 2018). Recent studies of biometrics in Africa, however, tend to emphasise the external nature (essentially non-African) of the biometric industry and its technologies. This chapter is driven by the need to recalibrate these findings and to reinscribe digital biometric systems into their terrain of production. How are the production relationships between public institutions, African software engineers, and foreign private suppliers organised and structured? What kind of social relations are the production relations of biometric systems? Why do these systems look so much alike and at the same time seem to mimic and reproduce the style of local government? What kinds of digital labour are implied by the efforts of public administrations to develop biometrics? Answering these questions first means recentring the production of the technology at the heart of African administrations, and thus to open a space for ethnographic observation, from a situated position, of both local and global relations of development. Second, it means considering production relations as a whole part of social relationships.

The chapter draws on three main sources: first, ethnographic observation of the workflows and production context of election technologies in Senegal and Kenya, carried out in Dakar (February–September 2017) and Nairobi (February–September 2019); second, interviews with executives, administrators, agents, and incumbent officials about the production of the technology; and third, interviews with biometrics industrial actors, vendors, and managers on and off site. The chapter reframes the African origins of biometrics (Breckenridge

2014), introducing new data on the organisation of digital labour and the role of voting in the emergence of global biometrics.

GLOBAL TECHNOLOGIES AND THEIR FIELD(WORK): EXTENDING THE STS DEBATE TO VOTING TECHNOLOGIES

Since its early days, STS has focused on how scientific knowledge, in addition to being a deeply local object, manages to travel with unique effectiveness to other contexts and spaces (Anderson and Adams 2008; Ophir and Shapin 1991; Shapin 1998). Through the study of the production of scientific facts (Star 1983), work contexts (Latour and Woolgar 1979), and devices/objects created (Latour n.b.), researchers have highlighted the role of scientists' work in making local knowledge a universal science. In so doing, they have sought to understand how people, through salaried, organised, and structured activities, successfully disconnect objects of knowledge from their immediate relations of production, escaping locality, and enabling them to integrate the social relations of other places and times (Akrich et al. 2011). As STS has globalised (Dumoulin Kervran et al. 2018), a growing number of studies have sought to understand how the Global South is redefining globalisation and technology. These terrains provide a vantage point to grasp the 'interconnected processes that drive science and technology' (Rottenburg, Schräpel, and Duclos 2012) that redefines globalisation as longstanding and informal circulations of people, goods, and markets (Choplin and Pliez 2015). Postcolonial studies of STS have argued that the (post)colonial geography and moment are conducive both to the genesis and development of scientific knowledge (Schiebinger 2005) and to the economic-industrial development of the former colonisers (Hecht 2004; Inikori 2002). Many Western sciences – especially medicine, biology, natural history, and botany – are now understood as sciences originating from a postcolonial context (Seth 2009). Other studies are rewriting the African history of technology (ceramics, ironwork, architecture, biometrics, nuclear energy, radiography, etc.), illustrating the shortcomings of the technology transfer narrative (Twagira 2020), reframing science and laboratories through open spaces – the forest, the plantation (Mavhunga 2014) – and thus restoring the roles of African actions, actors, institutions, and territories in innovation (Mavhunga 2017; Mika 2021; Osseo-Asare 2019).

Breckenridge wrote the South African history of biometrics technology and government (Breckenridge 2014, 2018). However, the contemporary literature on biometrics tends to portray it as a Western product, purchased and used by African countries following a pattern of technology transfer. Studies have investigated the diversion and appropriation of technologies by local actors (Beaudevin and Pordié 2016; Do Rosario and Muendane 2016), how controversies repoliticise them (Debos 2018; Salem 2018), how governments use them as state-building tools (Piccolino 2015; Rader and Périer 2017), the modernist imaginaries they convey (Cheeseman et al. 2019; Debos 2018), the injustices they produce (Amoah 2019; Eyenga et al. 2022), and how they affect electoral controversies (Emmanuel et al. 2019; Passanti 2021), reshaping the circulation of public knowledge on voting (Passanti and Pommerolle 2022). By focusing on the life of voting technologies after their production, this body of literature conveys a representation of biometrics as a ready-made product, the result of neutral production relationships occurring in an industry far removed from African soil and devoid of historicity.

Debos and Desgranges (2023) have shifted the debate to what takes place prior to voting, investigating the postcolonial dimensions of the biometrics market. The authors argue that European companies have positioned themselves in African markets since the 2000s, and that they maintain this position through political and economic business networks. In this chapter, I offer an alternative reading of how social relations are reproduced in and through technologies. This approach seeks to reintegrate the historical and contemporary African contribution to the biometrics industry—dating back to the 1970s—while acknowledging the North–South relations embedded within it. Drawing on lessons from both STS and the social history of elections on the objectification of social relations within technology (Garrigou 1993; Kelty 2008; Von Schnitzler 2013), the work of Pollock and colleagues (2007) on capturing the diversity of generic software packages, and that of Behrends and Rottenburg (2014) on travelling models, I propose the concept of division of biometric labour. The division of biometric labour consists of a formal – but not fixed – structure of work relationships between people and institutions coming from and holding together cross-cutting realities (North/South, production/use, private sector/ government) for the production of electoral biometrics. The concept posits social relations of work as a historically consolidated structure that enables their reproduction, and it offers a new perspective to rethink the (horizontal) globalisation of technology without overlooking (vertical) North-South relationships and the role of the South in innovation.

VOTING TECHNOLOGIES AND THEIR GENERIC BIOMETRIC CORE

Senegal and Kenya, located respectively at the western and eastern ends of Sub-Saharan Africa, do not share the same colonial history, do not face the same issues surrounding voting and government legitimacy, and organise elections in strikingly different ways. Indeed, Senegalese elections are managed by a directorate under the Ministry of Interior, while Kenyan elections are managed by an autonomous electoral commission. Beyond these differences, both countries manage voter identity through an identification system based on biometrics. Senegal's Ministry of Interior digitises citizen voter identification through a digital working infrastructure aimed at providing voter ID cards. When I conducted my fieldwork in 2019, the system was supplied by a Malaysian digital solutions company, IRIS Corporation Berhad, which had won the government contract. A local IT-services provider, Synapsys Conseils, worked with IRIS to ensure the day-to-day management of the technology. IRIS received the government contract money and subcontracted a portion of it to fund Synapsys' work. IRIS handled the delivery of equipment (computers/servers, smartcards, laser printers to engrave data on the cards, etc.), the troubleshooting of serious problems at server and printer level, and remote system support.

Synapsys was primarily responsible for designing the system, adapting it to local needs and language, developing new applications required to manage the electoral roll, designing the data entry interface, and training police officers in the use of the database. These activities take the form of a structured, institutional network of continuous daily work activities, carried out by police officers with varying degrees of digital skills, ranging from data entry clerks to IT engineers and managers (see Figure 10.1).

FIG. 10.1 The system administrator of the biometric citizen database of Senegal, Dakar, biometric ID card server room observation, Ministry of Interior, April 2022 (Cecilia Passanti)

The IT department of Kenya's electoral commission, on the other hand, has developed a network of tablets at polling stations connected with central data-centres, which allow for biometrically registering voters, electronically identifying them with their fingerprint before they vote, and transmitting election polling results to the central database for publication (see Figure 10.2). The voting process is almost entirely digital, with the exception of the voting itself, which is paper- and polling booth–based. During my fieldwork in 2017, the technology was provided by Safran Morpho, a well-known and longstanding French digital-identity solution provider. The IT directorate of the election commission was involved in daily technology production, managing the voter list database, recording equipment features and needs, maintaining the digital infrastructure for voting, training polling officials, and tallying up digital votes at the national tallying centres.

FIG. 10.2 Civic education video clip depicting two officials biometrically verifying the identity of a voter at a polling station, Nairobi, Kenya, May 2017 (Independent Electoral and Boundaries Commission)

As different as these technologies are, they share significant similarities, especially in terms of the organisation of the work involved in their production. Both rely on the work of a third party, a foreign technology provider specialising in biometric solutions, Safran Morpho in Kenya and IRIS in Senegal, that supplies the material infrastructure and delivers it on the ground. In both cases, after delivery, the technology is assembled, maintained, and produced in the field through a structured organisation of biometric work performed by public IT departments, managers, administrators, and thousands of civil agents involved in collecting, processing, and cleaning biometric data. These similarities suggest a formalised working structure that can manage and integrate the variety of different administrations.

Moreover, both technologies rely on a core structure of materials and software which are essential to any biometric system, namely the enrolment system, biometric scanners, the biometric software programs (AFIS), computers, and servers (see Figure 10.3). Some of these core components – especially the AFIS and the scanner to capture fingerprints – are mostly produced and sold by incumbent industrial manufacturers (such as Safran Morpho and IRIS). Vendors add on-demand modules to these core components based on local government requirements. For example, Senegal, which issues ID/voter cards, requires a 'printing system' that consists of a physical room in a protected environment, where civil servants print citizen data on biometric smart cards using biometric laser printers, and organise the delivery of the cards to citizens. Kenya, on the other hand, which transmits results digitally from the polling stations to the national tallying centre, requires an 'election result transmission system' that consists of an additional application embedded in the biometric tablet.

Biometric production relies on the design and formalisation of generic models of biometric identification systems that exist only on a theoretical level, on websites and at vendor booths. This generic structure is the result of the itinerant history of biometrics, guided by an economic model that aims to multiply markets to amortise high production costs by connecting to different organisational and geographic contexts. One of these was the African countries' quest, which occurred at the time of independence, for the material conditions of access to representative democracy. The administration of voting

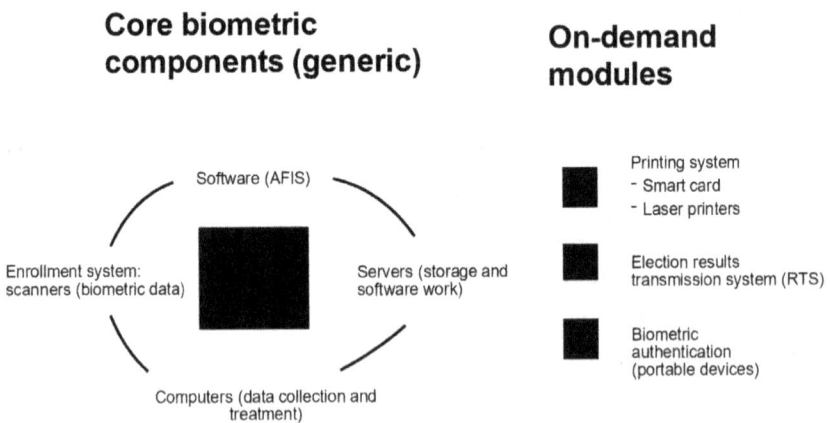

FIG. 10.3 A generic biometric identification system, which has a basic core to which the government can add modules upon request (Cecilia Passanti)

and governance, subject to the injunction of the development aid regime, was the terrain of several technoscientific inventions, including civil biometrics.

THE TRAVELLING HISTORY OF BIOMETRICS: GENERICNESS AND THE SOFTWARE'S AFRICAN TRAJECTORY

In the 1970s, fuelled by police departments' search for scientific methods to identify criminals, biometrics developed into a flourishing computer industry. Between the 1970s and the 1990s, the years during which African countries were achieving independence from colonial rule, biometrics were increasingly articulated with government identification systems in both the Global North and the Global South. Safran Morpho, one of the world's leading players in the industry, was created in the 1980s, under the name of Morpho Systèmes, by French investment funds and a research network based in Dakar (Senegal). The birth of Safran Morpho is closely linked to the history of access to independence and the spread of democracy in African countries.

In Africa, the colonial enterprise had imposed severe limitations on access to citizenship (Cooper 2012; Lonsdale 2016; Manby et al. 2011; Medina-Doménech 2009). This translated into a lack or non-existence of census systems for African citizens, and existing records were kept and maintained primarily in the metropole. The lack of registries became critical in the 1960s and 1970s, a period when the newly independent countries were to hold their first elections to establish their first independent government and thus establish their sovereignty over their people (Cooper 2012). In Senegal, the tension between the needs of the independent state and the absence of a solid identification system prompted the search for technical solutions, which was taken up by a group of engineers from the École des Mines in Paris, with the support of a French public bank (Caisse des dépôts et Consignations) and a French technology transfer provider working in French-speaking Africa (Sinorg). After the departure of the French administration, Sinorg computer scientists intervened within the ministries of the newly independent countries to support the administrations in the management of tasks, especially relating to taxation and the payment of public wages (Zimmermann 1984). The Sinorg experts, who were specialised in computer science for public administration, were also French-speaking African administration experts: they travelled from country to country to provide their technical support. They were thus able to recognise a cross-cutting issue faced by several African countries around the question of voter identification, which spurred the idea of investing in a technology company to solve it. The people involved in this early biometric project attested to the creation of Morpho Systèmes at the request of Senegal and more specifically Senegalese heads of state – first Léopold Sedar Senghor and then Abdou Diouf. Sinorg's CEO started hiring experts, especially French experts working with Senegalese computer scientists from public administration, to think about the use of biometrics and to constitute the voting list for the 1983 elections. Until then, biometric computing had been limited to criminal identification.

The shift of biometrics from forensics to civil services represented a real scientific and technical challenge. In its early days, biometrics was limited to police identification systems that worked with much smaller databases, containing only already-registered criminals and not the entire civilian population of a state. It was precisely to overcome this challenge that Morpho Systèmes was founded and heavily funded to research and produce a 'machine for African civil identification'. Using Senegalese fingerprints and databases, building on the 'African need' for voter identification, the first prototype of biometric voting – the Système Opérationnel de Saisie et d'Interrogations d'Empreintes (SOSIE) – was born in Dakar.

The electoral biometrics project in Senegal failed due to lack of sufficient funds. The government had disinvested and Sinorg's own investment was coming to an end, necessitating the signing of a new contract to continue the research. Neighbouring countries that had heard about the project, first and foremost Cameroon, funded the project for several years: in 1983, the government signed Morpho Systèmes' second contract to supply the first system (in the world) for producing national ID cards based on biometric identification of citizens. Traveling to Cameroon, the company had the opportunity to implement anew the working model on which the first Senegalese system was founded: an organisation of production that straddles the international supply of hardware and digital materials and a daily anchoring to the work of the local public administration. The production of ID cards is a long-term process, based on the daily demands of citizens, which takes place nationwide. Production is based on citizens' data collection, and it is not possible to produce it outside the field. Nor is it possible to produce it in a purely local way because much of the knowledge and materials on which it is based (in the same way as many information systems) is not present locally. The contract with Cameroon has provided Morpho with the necessary capital to continue research and development activities for several years. In addition, the establishment of a database of Cameroonian citizens has provided a development ground on which to improve the efficiency of the technology.

Besides these few early markets, the clients of biometrics for citizenship (voting and identity card production) were not enough to sustain research and development. Morpho thus returned to the forensic Western market, developing technical principles learned on the African field. The idea was to come back to the African market later on. The capacity acquired in Africa, however, had allowed Morpho to become a global leader – a position it still maintains. The company has signed contracts with other actors in other organisational settings and spaces, for example, two contracts for social security delivery in Florida and South Africa. Moreover, in the 1990s, the company was able to sign a contract to supply technology to the US FBI. The FBI market is the greatest reward for biometrics companies. Winning it means having a powerful and globally competitive algorithm in front of American and Asian companies. In each of these different contexts, the company has provided biometric identification systems through a production model situated between the global digital hardware market and the specific needs of local governments in different countries around the world.

The African context, especially that of Senegal, was mentioned by managers and engineers as the terrain that allowed them to think, mobilise sources, articulate needs, and experiment with the technique underlying the new technology. The Safran Morpho founder illustrated this role:

> The invention of biometrics for voter registration arose from a need expressed
> by Léopold Senghor, President of Senegal, and from SINORG's search for a
> solution. Personally, I believe it was a co-invention of Senghor – Senegal – and

the CEO of Sinorg. I was the one who made possible and realised what was only a dream.

The African contribution to the invention of electoral biometrics and of this new notion of biometrics for citizens (civil biometrics) is often stressed by Morpho founders and colleagues.

The birth phase of civilian biometrics occurred in the postcolonial era, as part of a research and development relationship between France and Senegal. The production relationships of the first prototype, traveling to various countries and contexts, became enduring and were structured into a production model that I now describe. It is based on a flexible structure of work activities that always remain the same but have a capacity to anchor themselves in the work of local administrations. Production relations keep the birth act of civil biometrics alive.

CONTEMPORARY RELATIONS OF PRODUCTION AS SOCIAL RELATIONS

As in the birth phase, the production relationships of biometric voter and citizen identification systems involve a set of work activities at the crossroads between national and transnational levels, local government and foreign technology providers. These relationships can be found in different countries, as I have observed in Kenya and Senegal, which suggests a travelling model of production. The production and implementation of biometric voter identification systems requires a set of work activities that I now disentangle.

The first type of activity relates to biometric software production. Software is often described as the beating heart of biometrics and as its most significant component because the deduplication of citizen files in the database depends on it. While software production is mostly an industrial activity, it does not take place outside the field. Biometric technology manufacturers secure their position by signing contracts, managing projects, and engaging in relationships with public administrations. Field experience is key to the development of biometric technology, as software production and performance rely on the availability of databases of real peoples' fingerprints. Moreover, variations in climatic and/or biological conditions dramatically influence the algorithms' performance. Predicting which kind of data can be obtained in the field is a very important form of expertise, as the Safran Morpho founder shows in the following interview excerpt:

> In a business there is technology and experience; technology without experience or experience without technology are worthless. For example, after a long plane ride, Bedouins during the dry season, Asian women and their small fingers make fingerprinting an extremely difficult and technical task. The software wonders if it has the characteristic points needed to calculate a score. If the fingerprints are very dry or small, the detection proves complicated and can lead to situations where you do not have enough characteristic points to compare the images. Once the problem is understood, solutions such as administering dry finger cream can be suggested.

Information about fingerprinting behaviour in a given population allows a company to know and learn about the efficiency and general behaviour of its own software. Each time a company signs a contract and implements its system in the field, it discovers different aspects of the efficiency of its algorithm in a specific context. This experience helps vendors overcome typical technical and administrative problems (Zviran and Erlich 2006) and thus supports the ability of biometric systems to better cater to an ever-wider range of social and environmental realities. Incumbent and longstanding software manufacturers have accumulated several more years of experience with African public administration, cultures, environments, and fingerprinting behaviours than other vendors. They are therefore more highly rated by African governments and in a better position to be chosen to carry out subsequent projects. Biometric software production and efficiency strongly relies on available datasets, and thus on citizens' personal data and on the ability of the company to be in the field.

Another set of production activities is referred to as system integration (SI). While incumbent technology providers can be or act as SI providers, the integration work can also be done by the administration's IT department and/or African IT companies. This is because being a system integrator involves dealing with oversight, the actual implementation of the technology in the field, and its material composition (integration for short), in order to make it a functional system. Integration is about knowing and cultivating relationships both with companies on the global biometrics market and with African administrations. These two types of relationships are referred to as 'partnerships' because they are built on mutual appreciation, trust, and win-win deals. Social relationships are the key feature of this marketplace: they are nurtured and cultivated through travel, visits, long-term friendships, knowledge exchange, advice, and trust. Power relations and hierarchies between actors need to be studied ethnographically at this level.

System integration involves three main activities. First, contract design consists of drafting specifications of technological requirements and qualities, for example, the number of devices required and their material shape. This activity is conducted in Africa by local IT departments, often in collaboration with international – often African – experts, and sometimes with technology vendors. The design and technical requirements of the system are established based on the election management software and products available. Second, the actual system integration occurs soon after the contract between the SI provider and the election institution or the government is signed. At this point, the SI provider calls on global partners (digital hardware and software vendors) to purchase the components they need in order to assemble an 'election system' (computers/servers, cameras, database management systems, biometric scanners, biometric printers, and biometric software). This horizontal network of peers allows vendors to purchase components from suppliers from around the world, to find the best price, or to form a long-term partnership. Third and lastly, implementation or adaptation to the local context consists in ensuring the effective operationality of technology that is customised to local needs and demand. Only once the contract is signed and the various parts have been purchased from the market is the generic technical design transformed into an existing technological object with extremely local qualities. The actual birth of a biometric election system only occurs at this stage, through an adaptation process that can take years. Adaptation involves adding the modules required by the contract, designing the technical system as such, adjusting data formats to other existing computer systems, adapting to local legislative requirements, tailoring machine programs to local programs, producing graphical interfaces for interaction with the database and biometric data (data entry and processing interface) accessible to officials (thus translating the system and interfaces into the language of the country), producing video demonstrations of the technology tailored to the local culture (with African rather than Caucasian actors, for example), and organising local management of distribution, shipping, and support. Incumbent software vendors have accumulated not only experience in the field of African public administration, but also partner relationships and knowledge of the global digital biometrics market. This allows them to always have the best prices on the market and therefore to continue to win contracts.

African public administrations are key actors in biometric production of technology systems and biometric data. This expertise materialises within a structured organisation of work. For example, since 1980, police officers with IT degrees attest their work with computers and databases. Since the 2000s, a private Senegalese technology provider named Synapsys Conseils has been working for the Senegalese Ministry of Interior for IT management of information systems. This makes the company a longstanding repository of knowledge on how public and administrative information systems work. The relationship between Synapsys and the ministry is a commercial one: the company is hired for a particular project, which it carries out with a group of employees that can vary in size. The relationship between the administration and the company is daily and informal, given their longstanding collaboration: the company has several offices where employees spend their days supporting the work of officials (police officers) whom they know well as people and employees. Similarly, in Kenya, the IT directorate of elections management employs a dozen IT staff that manage information systems before and after the arrival of biometrics.

The literature on contemporary voting technologies often treats local IT vendors and departments as intermediaries for bribes between the administration and foreign biometrics companies. However, these local actors are chiefly engaged in the day-to-day work of developing, implementing, and maintaining the technology. From an ethnographic perspective focused on observing daily activities, the work of the foreign companies is reduced to brief regulatory interactions when compared to the continuity of the relationships between administrators, managers, and machines that play out every day in administrations. These vignettes illustrate how foreign companies' interactions with African public administrations are intermittent and fleeting, and how they rely on local employment. They also highlight how production and maintenance relationships are most often peaceful relationships between human beings working together to make things work.

Nairobi, a few months before the 2017 elections

In the waiting room of the Kenyan Election Commission's IT director, I hear the voices of a group of men with marked French accents: a delegation from Safran Morpho has arrived in the field. Sitting on the same couch as me, a gentleman speedily types away on his laptop. I look at him several times until his gaze meets mine. 'Are you from Safran Morpho?', I ask him. He is the project manager of the French technology provider in charge of training local staff to use biometric tablets. He has to teach two main trainers, who will then go on to train both constituency-level IT support staff (140 people) and polling station staff (45,000 agents at voting offices). The company has rented a large shed on the outskirts of Nairobi and employed a hundred Kenyans to recharge the tablet batteries and put them inside the 45,000 backpacks to be sent to the voting offices. The company also employs a dozen Kenyan computer scientists, who represent the company on the ground. The delegation consists of three executives, directors, and the head of training. They will return to Paris next week after performing several demonstrations of the technology before election stakeholders.

Dakar, a few days before the 2021 elections

A delegation from Synapsys Conseils, the Malaysian company, is in Dakar in the field: a female director, two managers, and two technicians. The IT director of the public administration suggested that I meet with a member of the delegation at his office at the ministry. After a few minutes' wait, a very young Malaysian support engineer, tired from the Dakar heat and fasting for Ramadan, joins us. The two engineers can barely pronounce each other's names, but there is a strong shared understanding between them, certainly founded around computer science and perhaps their fasting fatigue. The Malaysian delegation is leaving today, after a nine-day stay in West Africa. The company normally intervenes remotely, but following a major technical problem with the ID-card servers in Guinea Conakry, right on the border with Senegal, the team took the opportunity to come to Dakar to solve minor problems and speak with the authorities.

Foreign companies' interventions in the field rely on quick actions by small delegations. Beyond the sending of materials, interventions can be technical (technology training, troubleshooting), organisational (recharging and distribution of machines, public demonstrations), or political (company executives interacting with the leaders of local institutions). These interventions consist of social relationships aimed at establishing a functional biometric system but also a market relationship, based on trust and long-term partnerships between the company and the administration. The speed of projects is counterweighed by the need to maintain a strong presence on the ground, to ensure the continuity of the daily support that technical systems require in order to function. My intention, in emphasising the work of local IT, is not to minimise the work of biometrics companies surrounding product ideation, supply, shipping, transfer, and so forth. However, as they are not based in African countries, the companies work remotely and rely on local IT organisations to implement projects.

Last but not least, biometric voter identification systems require the extensive mobilisation of human resources by local governments. In both Senegal and Kenya, the number of civil servants required to collect, verify, process, and transmit biometric data is impressive and organised into workflows, starting with enrolment to register voters and process requests for ID documents. While the directorate overseeing elections in Senegal relies on the permanent employment of police agents as well as some temporary agents, the Election Commission of Kenya employs temporary agents to enrol voters several months before an election takes place, in order to be able to use the biometric tablets to identify voters on the polling date and to send the results for tallying. During elections, these groups of people, numbering in the thousands – 400,000 in Kenya and

100,000 in Senegal – register and verify millions of voters across the country – 15 million in Kenya and 17 million in Senegal – for several hours a day and over several months. This group of people is framed as users of biometric technologies. However, this label does not aptly describe the capital role of their work in the production of biometrics. During voter registration, civil servants actually produce the main component required for the identifier systems to function: the biometric data. Other teams of civil servants (in Senegal mostly women hired on short-term contracts) use computers to cleanse the data collected in the field and prepare them for inputting into the database. Before the data are actually entered into the database, the biometric software compares each incoming file with those already registered, validates unique requests, and blocks duplicate files by suspending them for verification by a human agent.

A team of police investigators is assigned to verify suspicious files. By analysing the data shown on the biometric software users' interface, investigators work to clarify the reasons why a citizen has made a duplicate ID request. The biometric software has automated and thus reduced some of the work carried out by police investigators to identify suspicious cases, but this work is not replaceable. The more suspicious cases the software identifies, the more human work is required to verify the suspicious files. The fewer cases it identifies, the less human work is required, but the more the system will issue identities to people requesting an ID card without them actually having the credentials required. After the citizen files have been processed by the software and the police investigators, they are sent for printing, a process referred to as the printing and distribution system. This system is managed by a team of police officers who work with computers, smart cards, and biometric laser printers within a printing room organised into several laboratories. Groups of officers specialise in one function and perform their respective roles, passing the baton to the next group once they have completed their tasks: launching the print files from the central computer; confirming the printing jobs; inserting blank cards into the printers; collecting the printed cards; verifying the data on each card against the computer's digital data; packaging a batch of cards; and dispatching the delivery to the administration on the ground. While card production volumes normally amount to 2,000 cards per day, a few months before the election, the demand for documents begins to increase, and productivity becomes a real democratic value – the more cards are delivered, the more people will be able to vote – and card production reaches volumes ranging between 40,000 and 70,000 cards per day.

The analysis of production relations attests that at each stage of technology production, actors have social and collaborative relationship at the crossroads between national and transnational levels, local government, and foreign technology providers. Labour relationships serve as a catalyst for such diverse realities as government and technology providers, the nation-state and the global digital marketplace, the public and private sectors, and the South and the North. These relationships hold together complex artefacts such as ID systems and enable them to travel. While the North-South relation can be very important in the production of electoral biometrics, it is not necessary. It can morph into Asia-Africa relations, as is the case in Senegal, and, depending on government choices, it can morph into an African government that resorts independently in the global digital market. What really matters and cannot be replaced is the working model that links the global approvisioning market for digital materials to the biometric work of public administration. This working model can be found in different countries and settings and is summarised in Figure 10.4.

CONCLUSION

The African trajectory of global biometrics and the contemporary relations of production of biometric-based voting technology offer a new perspective on the ability of this technology to spread around the world.

Biometric technologies – both electoral and generic – appear as open, constantly evolving systems that have specialised in dealing with different organisations and fields. This specialisation has been integrated within the structure of production. In the two countries where I observed them, biometric-based voting technologies are built on a complex organisation of work that involves the daily labour of both national and transnational institutions and people. These relations of production are organised in a formal structure that I call the division of biometric labour. This structure was born in Africa, particularly Senegal, at the time of independence, through a network of North-South (France-Senegal)

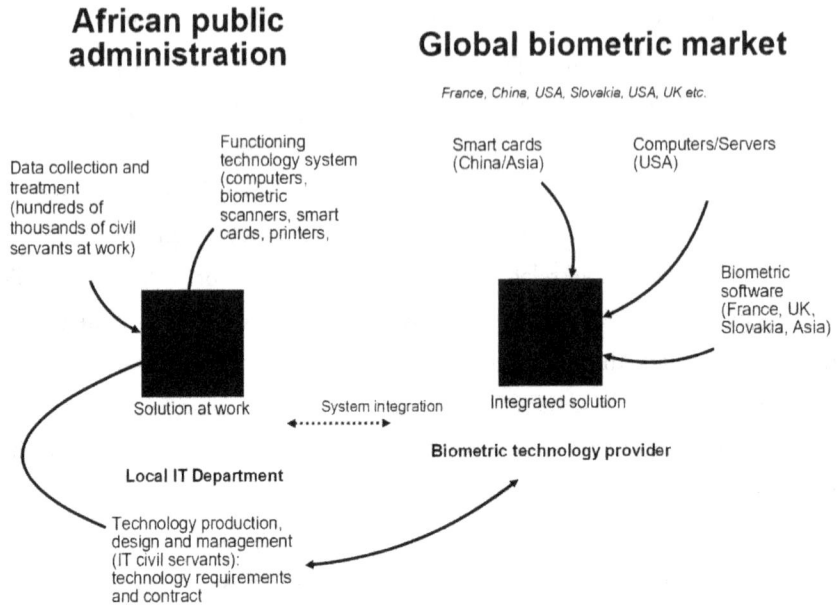

FIG. 10.4 The division of biometrics labour between African public administration and the biometric market (Cecilia Passanti)

research and development relations. The history of electoral biometrics explains why these technologies are so adaptable to African democracies and to post-colonial countries that share a similar history surrounding civil identification. Thus, not only has African countries' quest to ensure the material conditions of democracy played a key role in the creation and spread of biometrics in the Global South, but it has also played a key role in the growth of the industrial and global biometrics project itself, pushing biometrics out of the forensic sector and into the civilian sector.

The chapter documents an aspect that is often underestimated in studies of globalisation as it is too often understood only through the notion of 'value chain': labour relations in the fabrication of technologies cannot be reduced to unfair labour organisation. Following that line of inquiry, the chapter shows another dimension of the globalisation of technology as experienced from Global South countries, one which relates to role circulation and market expansion through work relations. In biometric science, technology, and industry, genericness and globalisation are not achieved through detachment work from the original social relations of production. On the contrary, genericness and globalisation are achieved through the original act of invention, based on the North-South relationship of innovation (the 'birth stage' [Pollock 2007]). Moreover, they are achieved through the reiteration of these original social relations of production over different places and spaces. The reiteration of production relations allows for their specialisation and structuring; it enables them to objectify into a hard core of technology production relation. These production relationships make biometric technologies, which are vast technologies based on large-scale administrative infrastructures, travel with a 'unique efficiency' (Shapin 1998) and become a global presence.

REFERENCES

Akrich, M., M. Callon, B. Latour, and A. Monaghan, 'The Key to Success in Innovation Part I: The Art of Interessement', *International Journal of Innovation Management*, 6.2 (2011): 187–206 https://doi.org/10.1142/S1363919602000550.

Amoah, M., 'Sleight Is Right: Cyber Control as a New Battleground for African Elections', *African Affairs*, 119.474 (2019): 68–89 https://doi.org/10.1093/afraf/adz023.

Anderson, W., and V. Adams, 'Pramoedya's Chickens: Postcolonial Studies of Technoscience', in E. Hackett, O. Amsterdamska, M. Lynch, and J. Wajcman, eds., *The Handbook of Science and Technology Studies* (MIT Press, 2008), pp. 181–204.

Beaudevin, C., and L. Pordié, 'Diversion and Globalization in Biomedical Technologies', *Medical Anthropology*, 35.1 (2016): 1–4 https://doi.org/10.1080/01459740.2015.1090436.

Behrends, A., S.-J. Park, and R. Rottenburg, *Travelling Models in African Conflict Management: Translating Technologies of Social Ordering* (Brill, 2014).

Bierschenk, T., and J.-P. Olivier de Sardan, 'How to Study Bureaucracies Ethnographically?', *Critique of Anthropology*, 39.2 (2019): 243–57 https://doi.org/10.1177/0308275X19842918.

Breckenridge, K., *Biometric State: The Global Politics of Identification of Surveillance in South Africa, 1850 to the Present* (Cambridge University Press, 2014).

Breckenridge, K., 'État documentaire et identification mathématique: La dimension théorique du gouvernement biométrique africain', *Politique africaine*, 152.4 (2018): 31–49 https://www.cairn.info/revue-politique-africaine-2018-4-page-31.htm?contenu=resume.

Breckenridge, K., and S. Szreter, eds., *Registration and Recognition: Documenting the Person in World History* Oxford: Oxford University Press, 2012).

Cheeseman, N., G. Lynch, and J. Willis, 'Digital Dilemmas: The Unintended Consequences of Election Technology', *Democratization*, 25.8 (2018): 1397–1418 https://doi.org/10.1080/13510347.2018.1470165.

Choplin, A., and O. Pliez, 'The Inconspicuous Spaces of Globalization', *Articulo – Journal of Urban Research*, 12 (2015) https://doi.org/10.4000/articulo.2905.

Cole, S., *Suspect Identities: A History of Fingerprinting and Criminal Identification* (Harvard University Press, 2009).

Cooper, F., 'Voting Welfare and Registration: The Strange Fate of the État Civil in French Africa, 1945–1960', in K. Breckenridge and S. Szreter, eds., *Registration and Recognition: Documenting the Person in World History* (Oxford University Press, 2012), pp. 385–412.

Debos, M., 'La biométrie électorale au Tchad: Controverses technopolitiques et imaginaires de la modernité', *Politique africaine*, 152.4 (2018): 101–20 https://www.cairn.info/revue-politique-africaine-2018-4-page-101.htm.

Debos, M., and G. Desgranges, 'L'invention d'un marché: Économie politique de la biométrie électorale en Afrique', *Critique internationale*, 98.1 (2023): 117–39 https://www.cairn-int.info/journal-critique-internationale-2023-1-page-117.htm.

Do Rosario, D. M., and E. E. Muendane, '"To Be Registered? Yes. But Voting?" – Hidden Electoral Disenfranchisement of the Registration System in the 2014 Elections in Mozambique', *Politique Africaine*, 144.4 (2016): 73–94 https://www.cairn-int.info/journal-politique-africaine-2016-4-page-73.htm.

Donovan, K., 'The Biometric Imaginary: Bureaucratic Technopolitics in Post-Apartheid Welfare', *Journal of Southern African Studies*, 41.4 (2015): 815–33 https://doi.org/10.1080/03057070.2015.1049485.

Dumoulin Kervran, D., M. Kleiche-Dray, and M. Quet, 'Going South: How STS Could Think Science in and with the South?', *Tapuya: Latin American Science, Technology and Society*, 1.1 (2018): 280–305 https://doi.org/10.1080/25729861.2018.1550186.

Emmanuel, D., E. John, and I. Owusu-Mensah, 'Does the Use of a Biometric System Guarantee an Acceptable Election's Outcome? Evidence from Ghana's 2012 Election', *African Studies*, 78.3 (2019): 347–69 https://doi.org/10.1080/00020184.2018.1519335.

Eyenga, G. M., G. O. Omgba, and J. F. Bindzi, 'Être sans-papier chez soi? Les mésaventures de l'encartement biométrique au Cameroun', *Critique Internationale*, 4.97 (2022): 113–34 https://www.cairn.info/revue-critique-internationale-2022-4-page-113.htm.

Garrigou, A., 'La construction sociale du vote: Fétichisme et raison instrumentale', *Politix. Revue des Sciences Sociales du Politique*, 6.22 (1993): 5–42 https://www.persee.fr/doc/polix_0295-2319_1993_num_6_22_2042.

Hecht, G., 'Colonial Networks of Power: The Far Reaches of Systems', *Annales Historiques de l'Electricité*, 2.1 (2004): 147–57 https://www.cairn.info/revue-annales-historiques-de-l-electricite-2004-1-page-147.htm.

Inikori, J. E., *Africans and the Industrial Revolution in England: A Study in International Trade and Economic Development* (Cambridge University Press, 2002).

Kelty, C., 'Towards an Anthropology of Deliberation', Society for Social Studies of Science Annual Meeting, Montreal, 2008 http://kelty.org/or/papers/unpublishable/Kelty_Anthro_of_Delib_2008.pdf.

Latour, B., 'Notes sur certains objets chevelus', *Nouvelle Revue d'Ethnopsychiatrie*, 27 (1995): 21–33 http://www.bruno-latour.fr/fr/node/228.html.

Latour, B., and S. Woolgar, *Laboratory Life: The Construction of Scientific Facts* (Princeton University Press, 1979).

Lonsdale, J., 'Unhelpful Pasts and a Provisional Present', in E. Hunter, ed., *Citizenship, Belonging, and Political Community in Africa: Dialogues Between Past and Present* (Ohio University Press, 2016), pp. 17–40.

Manby, B., H. Bazin, T. Vedel, and C. Becker, *La Nationalité en Afrique* (Karthala, 2011).

Mavhunga, C. C., *Transient Workspaces: Technologies of Everyday Innovation in Zimbabwe* (MIT Press, 2014).

Mavhunga, C. C., *What Do Science, Technology, and Innovation Mean from Africa?* (MIT Press, 2017).

McLaughlin, J., P. Rosen, D. Skinner, and A. Webster, *Valuing Technology: Organisations, Culture and Change* (Routledge, 2002).

Medina-Doménech, R., 'Scientific Technologies of National Identity as Colonial Legacies: Extracting the Spanish Nation from Equatorial Guinea', *Social Studies of Science*, 39.1 (2009): 81–112 https://doi.org/10.1177/03063127080976.

Mika, M., *Africanizing Oncology: Creativity, Crisis, and Cancer in Uganda* (Ohio University Press, 2021).

Ophir, A., and S. Shapin, 'The Place of Knowledge: A Methodological Survey', *Science in Context*, 4.1 (1991): 3–22 https://doi.org/10.1017/S0269889700000132.

Osseo-Asare, A. D., *Atomic Junction: Nuclear Power in Africa After Independence* (Cambridge University Press, 2019).

Passanti, C., 'Contesting the Electoral Register During the 2019 Elections in Senegal: Why Allegations of Fraud Did Not End with the Introduction of Biometrics', *Jan Thorbecke Verlag*, 48 (2021): 515–25 https://doi.org/10.11588/fr.2021.1.93969.

Passanti, C., and M.-E. Pommerolle, 'The (Un)Making of Electoral Transparency Through Technology: The 2017 Kenyan Presidential Controversy', *Social Studies of Science*, 52.6 (2022): 928–53 https://doi.org/10.1177/0306312722112400.

Piccolino, G., 'Infrastructural State Capacity for Democratization? Voter Registration and Identification in Côte d'Ivoire and Ghana Compared', *Democratization*, 23.3 (2015): 1–22 https://doi.org/10.1080/13510347.2014.983906.

Pollock, N., R. Williams, and L. D'Adderio, 'Global Software and Its Provenance: Generification Work in the Production of Organizational Software Packages', *Social Studies of Science*, 37.2 (2007): 254–80 https://doi.org/10.1177/030631270606602.

Rader, A., and M. Périer, 'Politiques de la reconnaissance et de l'origine contrôlée: La construction du Somaliland à travers ses cartes d'électeurs', *Politique Africaine*, 4.144 (2017): 51–71 https://www.cairn.info/revue-politique-africaine-2016-4-page-51.htm.

Rottenburg, R., N. Schräpel, and V. Duclos, 'Relocating Science and Technology: Global Knowledge, Traveling Technologies, and Postcolonialism. Perspectives on Science and Technology Studies in the Global South', Max-Planck-Institut für Ethnologische Forschung, 19–20 July 2012 https://www.hsozkult.de/event/id/event-68318.

Salem, Z. O. A., '"Touche pas à ma nationalité": Enrôlement biométrique et controverses sur l'identification en Mauritanie', *Politique Africaine*, 152.4 (2018): 77–99 https://www.cairn.info/revue-politique-africaine-2018-4-page-77.htm.

Schiebinger, L., 'Forum Introduction: The European Colonial Science Complex', *Isis*, 96.1 (2005): 52–55 https://doi.org/10.1086/430677.

Seth, S., 'Putting Knowledge in Its Place: Science, Colonialism, and the Postcolonial', *Postcolonial Studies*, 12.4 (2009): 373–88 https://doi.org/10.1080/13688790903350633.

Shapin, S., 'Placing the View from Nowhere: Historical and Sociological Problems in the Location of Science', *Transactions of the Institute of British Geographers*, 23.1 (1998): 5–12 https://doi.org/10.1111/j.0020-2754.1998.00005.x.

Star, S. L., 'Simplification in Scientific Work: An Example from Neuroscience Research', *Social Studies of Science*, 13.2 (1983): 205–28 https://doi.org/10.1177/030631283013002002.

Star, S. L., and K. Rudleder, 'Steps Toward an Ecology of Infrastructure: Design and Access for Large Information Spaces', *Information Systems Research*, 7.1 (1996): 111–34 https://doi.org/10.1287/isre.7.1.111.

Twagira, L. A., 'Introduction: Africanizing the History of Technology', *Technology and Culture*, 61.2 (2020): S1–S19 https://doi.org/10.1353/tech.2020.0068.

Von Schnitzler, A., 'Traveling Technologies: Infrastructure, Ethical Regimes, and the Materiality of Politics in South Africa', *Cultural Anthropology*, 28.4 (2013): 670–93 https://doi.org/10.1111/cuan.12032.

Zimmermann, J.-B., 'Politiques africaines de l'informatique', *Politique Africaine*, 13 (1984): 79–90 https://www.persee.fr/doc/polaf_0244-7827_1984_num_13_1_3688.

Zviran, M., and Z. Erlich, 'Identification and Authentication: Technology and Implementation Issues', *Communications of the Association for Information Systems*, 17.4 (2006): n.p.

11

HOW MAGIC BULLETS TRAVEL: AN ACCOUNT OF READY-TO-USE THERAPEUTIC FOOD IN INDIA

Aamod Utpal

POSTCOLONIAL STUDIES OF TECHNOSCIENCE GENERALLY FOCUS ON TECH-nologies that circulate, change directions, and reverse their flows (Anderson 2002). They articulate a critique of the idea of universal and immutable technologies that would exist similarly in both the Global North and the Global South. However, certain technologies are emphasised as ends in themselves, circulating mostly one way – so much so that they seem to keep contesting the circular imagination of technical objects. A demonstrable example of such a technology is the lifestraw, a portable straw with filtering capabilities that can be used in any water source (rivers, ponds) to filter clean water, which was conceptualised and designed by a European company as a humanitarian good destined to end up in crisis situations in Africa (Redfield 2016). Humanitarian technologies belong to an original category of transnational commodities travelling through circuits of unequal exchange. They circulate in spaces dominated by moral economies and ethical considerations surrounding the wellbeing of 'distant others' (Redfield 2012) – the 'other' being the recipient of humanitarian aid in the Global South (Fassin 2011; Redfield 2012). Such technologies are designed to be used at a point of reception, and care, that is deeply merged in the identity of sites defined as underdeveloped, poor, in crisis – in other words, non-North. In this sense, humanitarian exchanges are unequal and unilateral by design, even if bottom-of-the-pyramid approaches geared towards finding market-based solutions do exist.

In spite of this deeply anchored identification of points of care, humanitarian logic mostly expects humanitarian technologies to be context agnostic, and their use value to be equivalent to their materiality (Scott-Smith 2013, 2016). As Collier et al. (2017) illustrate, such humanitarian life technologies are designed as small-scale and minimalist palliatives, as is the case of the lifestraw. Hence, these 'little development devices', as Collier and colleagues (2017) call them, can also be thought of as demonstrating a rather short-lived sociotechnical imaginary: humanitarian technologies do not promise food or health security for vulnerable populations, merely the survivability of children and vulnerable populations till the end of the current crisis. The stakeholders at the receiving end of such transcontinental humanitarian efforts are often disenfranchised, 'geographically marginal' (Collier et al. 2017), displaced people and weak or barely existent local social security systems. Premised on such a skewed power dynamic, the dichotomy between the innovative (and charitable – as humanitarian efforts are often charity funded) Northern actors and the Southern 'beneficiaries', between order/chaos and formal/informal, is inescapable.

Contrary to this view of technologies as context agnostic and overdetermined by use value, anthropologists of technical objects dispute the notion that the value of technical objects is inherent and fixed, but claim that along their social life, their value is dynamic and multiple (Appadurai 1988). As the object passes through its biography, it leaves and enters different phases or regimes of valuation, where different values are ascribed to it in relation to its sociocultural milieu (Appadurai 1988; van der Geest 1996). This concept has most notably been applied to pharmaceuticals to account for the fluidity of meanings attached to them in different situations, which has ultimately helped anthropologists understand why and how they work (or don't) in certain situations (Ecks and Basu 2009; Hardon and Sanabria 2017; Whyte et al. 2002). The adaptability of such technologies has even been defined as a normative feature, as in the famous case study of the Zimbabwean bush pump (de Laet and Mol 2000), that shifted attention to the multiple ways in which global technologies create local, fluid contact zones (Anderson 2002), where actors of the Global South interact with the technologies meaningfully and creatively (de Laet and Mol 2000; Redfield 2016). But how definitive is this opposition between unchangeable technologies and fluid ones? And how can stability be performed through the multiplicity of appropriations? One way of answering is to look at the actual circulations and discussions surrounding technological movability.

In order to carry this discussion forward, in this chapter I critically examine the case of RUTF's reception in India. Ready-to-use therapeutic food (RUTF henceforth) is a peanut- and milk-based paste used in humanitarian emergencies to save children suffering from severe acute malnutrition (SAM) from early mortality. In order to answer questions about its success as a humanitarian commodity, its material characteristics have often been analysed (Scott-Smith 2018). On the contrary, a critical approach to RUTF consists in looking at it in relation to the multiplicity of values attached to it, not restricted to its medicalised nutritional ambition. This will lead us to ask: What are the ways in which actors in the Global South question, resist, and appropriate the under-lying premise of a transcontinental humanitarian 'exchange' in the context of RUTF and SAM?

In the following pages, I first try to establish why RUTF can be seen as emblematic of technoscientific expansion into issues of acute food insecurity, and consequently, what it means for the kinds of production models and exchanges it engenders and the specific markets it shapes. I also present the general scope within which humanitarian solutions for SAM are expected to operate and the rationale underlying these solutions. The chapter then revisits these encounters between the object and the humanitarian subject from the vantage point of actors in India. I organise the remainder of the chapter into three sections that describe the way in which different Indian stakeholders – civil society, the private sector, and the regulatory authorities – have engaged with the idea of RUTF. Ultimately, the chapter illuminates the fundamentally unequal premise of a contemporary humanitarian exchange, one that assumes passive Southern actors as recipients of technologically constituted globalised solutions from the North in the context of weak local institutions. I show that when confronted with RUTF, which is treated as a single-value technical product, the stakeholders in India question this aforementioned premise by employing their own unique valuing strategies to frame and value RUTF in a way that goes beyond its utilitarian imagination.

To trace the journey of RUTF in India, I draw on secondary data from available secondary literature, newspaper reports, publicly available government policy papers, publicly available company websites, grey literature, and regulatory texts; and I support this with primary data based on five in-depth interviews with stakeholders involved in discourse on SAM management, especially in India.[1]

RUTF AS A MAGIC BULLET IN CHANGING HUMANITARIANISM

Severe acute malnutrition is identified as one of the leading causes of early childhood mortality worldwide[2]. Children who are diagnosed as suffering from SAM are reported to be as much as nine times more vulnerable to early mortality (Black et al. 2008). SAM is most common in emergency contexts, such as civilian emergencies (war and refugee crises), climate disasters, and acute crop failures. This condition primarily affects children (below six years of age) and can materialise in two ways: wasting (of body muscles – formerly called marasmus); and generalised oedema/swelling (previously referred to as kwashiorkor). Both can also be observed at the same time (Grover and Ee 2009). Currently, the scientific consensus is that SAM is multifactorial by nature; it is caused by a synergism between acute lack of food and childhood infections (Black et al. 2008).

Before the advent of RUTF and the Community-Based Management of Acute Malnutrition (CMAM) program, the treatment of SAM usually was clinic-based (Scott-Smith 2013). Children with SAM usually presented with heavy weight loss, loss of appetite, or oedema (swelling) (Grover and Ee 2009). These children were admitted to an in-patient facility either at an existing health establishment or, in case of infrastructure breakdown, at a makeshift treatment centre. The treatment protocol involved medical attention for complicated SAM cases (complicated cases here correspond to children with loss of appetite, a baggy-pants appearance, or bipedal oedema), along with a regimen of therapeutic milk (F-100 and F-75) for weight gain, after which the children were discharged into the community and put on home-based diets or certain protein supplements like corn-soy blend (Khanam et al. 1994).

The early 2000s saw famines in DRC, Sudan, Ethiopia, and Niger: this is where RUTF was introduced extensively by aid agencies for SAM management. As the story goes, it was first developed in a domestic blender by a paediatrician working for the IRD (Institut de Recherche pour le Développement, France), who improved upon the previous milk-based formula by replacing part of the skimmed milk with peanut butter paste[3]. A flurry of research activity around this promising product led to excellent results in field trials in Malawi, Chad, and Ethiopia (Collins and Sadler 2002; Diop et al. 2003; Manary et al. 2004). The product is composed of milk solids, peanut paste, and carbohydrates fortified with a mineral premix. It is an energy-dense high protein paste (per 100 gm paste, 520–550 kilocalories and 13–15 gm of protein). Packaged as a homogenous paste in 92-gm sachets, this product, combined with a simplified regimen of early identification and classification of children affected with SAM, showed significant improvement over other alternatives in solving the issues faced previously by these organisations when trying to organise efforts to cure or minimise damage due to SAM on scale (Briend and Collins 2010; Scott-Smith 2018). The key properties of RUTF as compared to other products were (a) lower water activity than older formulae used to treat SAM, such as F-100 therapeutic milk, leading to longer storage times and less chance of spoilage and subsequent infection (Manary et al. 2004); (b) higher energy density than older alternatives like F-100 and corn-soy blends, thus resulting in higher and quicker average weight gain in SAM children and improved recovery rates; (c) it needs no cooking but can be distributed to resource-constrained families directly; and (d) it can be used in the community, away from the costly setups of inpatient/facility-based treatment, and it can be rapidly deployed in emergency situations (Collins and Sadler 2002). The apparent success of RUTF in community feeding programmes (CMAM)[4] helped launch humanitarian efforts at a bigger scale than before. More generally, it gradually became emblematic of deeper mutations in humanitarianism during this time. Let us now turn to these changes.

In the 1970s and 1980s, large-scale agrarian projects led by agronomists and national planners focused on improving farming practices in the South to improve food security (Jézéquel 2015; Ruxin 1996). Usually, these projects would be integrated with national planning goals and have cross-cutting horizontal emphasis. They were later replaced by projects more modest but technologically constituted in their goals. Helmed by different categories of experts such as nutritionists and doctors, these projects would favour vertical programs with tangible, measurable impact (Jézéquel 2015; Redfield 2017; Ruxin 1996; Scott-Smith 2020). The introduction of RUTF brought about a further narrowing and technologisation of malnutrition management programmes, while massively increasing the scale of operations. This tendency to promote unique products for modest survival goals has also been articulated by invoking another term – 'magic bullets'.

'Magic bullet' in immunology refers to 'drugs that go straight to their intended cell-structured targets' (Strebhartdt and Ullrich 2008). In development projects, magic bullets are technical interventions aimed at quick, minimalist impacts, making them appealing for universal application in varied humanitarian contexts (Mason and Margetts 2017; Redfield 2017). The magic bullet analogy is further bolstered by the fact that RUTF is also understood as a vessel or 'tool for delivering improved nutrition', owing to its lipid paste being able to preserve micronutrient premixes well.

> With the concept of paste, it was basically a very good tool to deliver improved nutrition. You could improve diet by putting essential elements that were missing in the diet in a novel way. (Interview with NGO representative)

RUTF exemplifies this concept by offering a standardised pharmaceutical solution to malnutrition, reflecting a shift towards medicalised and industrial food practices (Shukla and Marathe 2017).

The idea of RUTF as a magic bullet has been used to emphasise its insistence on this specificity of objective (to save the child suffering from SAM from mortality) and its technologisation/medicalisation of food insecurity. The concept of a magic bullet has also been mobilised for its symbolic purpose. If a metaphorical bullet is discharged, the assumption is that it must end up in its metaphorical target. By definition, its value lies in its efficacy and the minimalist, straight path it traverses. In the next section, we therefore examine this path to see how this 'magic bullet by destination' gets valued by stakeholders in India – and how its trajectory might be diverted.

A PANOPLY OF STAKEHOLDERS — INDIA, RUTF, AND VALUATION STRATEGIES

In September 2017, a newspaper reported that India's Ministry of Women and Child Development had written to the respective states cautioning against RUTF on the grounds that that there was a lack of evidence regarding the usage of RUTF and that its usage had the potential to impact 'nutritional best practices' and continued breastfeeding among children older than 6 months[5]. In November of the same year, another newspaper reported a change in the policy stipulating that the decision to procure RUTF should be left with the discretion of individual states in India[6]. Furthermore, newspapers reported that, responding to a petition filed in the high court of Maharashtra (a state in western India), the Maharashtra state government informed that it was stopping the procurement process of RUTF[7][8]. In 2019, that state finally procured RUTF (now named EDNF – energy dense nutrition food) for large-scale deployment in a few districts[9]. This was again opposed by some civil society organizations.

As technological objects travel through their circuits, they are valued differently at different points and spaces by actors for their own respective purposes. Owing to India's dual role as a site for both production and consumption, it assimilates within its confines a heterogeneity of valuation strategies and actors that are entangled in various ways, often leading to conflicting strategies. This entanglement allows us to present a contrast to how magic bullets are otherwise assumed to work in a humanitarian exchange, as a context agnostic technical projectile targeting a specific, narrowly defined problem. The following section expands on this tension that India's position as a supplier and consumer of humanitarian goods and services produces to understand the stakeholders involved and their different valuation strategies.

United Nations Children's Fund (UNICEF) and CODEX[10] guidelines clearly specify the proportions of protein, energy, and micronutrients in RUTF, per kilogram of body weight and per day. RUTF is industrially prepared according to Good Manufacturing Practices (GMP) standards, and nearly 80% of it is procured by UNICEF from a list of accredited manufacturers based in Asia, Africa, Europe, and North America. RUTF was originally only manufactured under patent protection (brand name Plumpy'Nut) in France by the French company Nutriset. The company, under its PlumpyField network arrangement, had also tried to foster local manufacturing in countries where it was deployed. Since then, the number of manufacturers has expanded. Table 2 shows the manufacturers based on their locations; 5 of 21 of these accredited RUTF manufacturers were based in India until 2021[11], which is the highest for any individual country in this list.

CONTINENT	TOTAL NO. OF PRODUCERS	OF TOTAL, NO. OF INDEPENDENT MANUFACTURERS	NO. OF PLUMPYFIELD NETWORK MEMBERS
Americas	3	1	2
Europe	1	0	1
Africa	11	5	6
Asia	6	5	1

TABLE 2 Number of UNICEF suppliers of RUTF and their locations, 2023 (UNICEF RUTF market update 2023)

In terms of gross value, the largest UNICEF order of all types of goods and services, in 2020, was sourced from India[12]. India was also the third largest procurement source in 2021 – it was the country with the highest number of vendors invited by UNICEF to bid for its procurement tenders[13]. Within the humanitarian supply chain organised by UNICEF, India clearly features prominently as an important site of production. In 2021, India was the sixth largest destination for all commodities procured from UNICEF (in gross value in US$)[14]. This positions India as an important site for consumption as well.

India's high malnutrition numbers attracted attention from international humanitarian organisations, and it was in the aftermath of a devastating flood in the eastern state of Bihar in 2007 that RUTF was first introduced in the country by Médecins Sans Frontières (MSF). However, a sequence of events later, the Indian government had stopped another organisation, UNICEF, from conducting a similar intervention elsewhere in India, which included sending back shipments of imported RUTF. What caused such contrasting responses?

In MSF's case study of humanitarian negotiations, it outlines the conditions on the ground in 2008–2009, before the RUTF was about to be introduced in India. The report shows that a diverse range of stakeholders, including civil society groups like the Right to Food Campaign, central and state governments, and the Indian legal system (the High Court and the Supreme Court), were debating how best to cure children with malnutrition. MSF was committed to treating children suffering from SAM by first making the problem 'visible' (interview with NGO representative) using simplified SAM detection tools like the mid-upper arm circumference (MUAC) tape, followed by a quick intervention: 'the priority is to treat the children and if their treatment is not to be delayed we have to use whatever therapeutic products are available' (Doyon 2011). The priority was to act 'immediately and locally', owing to the precarity of children suffering from SAM, and to avoid what was understood by certain sections of MSF as 'procedural straitjackets' in the host country. RUTF was eventually introduced in an intervention in Bihar by MSF, responding to a devastating flood there. The difference in opinion with the Right to Food Campaign members was on the applicability of RUTF and what it represented. For MSF, RUTF was necessary in SAM mitigation programmes because it simplified distribution and use, mainly because under its community-based intervention, the 'medical responsibility was to be delegated to the families of malnourished children' (Doyon 2011). This individualised medical regime that prioritised quick interventions and minimalist procedures over bureaucratic delays was opposed by the civil society members, who saw it as representative of a commercial interest in a 'common good' (Doyon 2011).

Another manifestation of these contradictory valuation strategies happened when UNICEF introduced this intervention as an extension of its earlier institutional programme of treating SAM in Madhya Pradesh in 2008. This intervention was later struck down by the central Indian government, which was acting according to a previous Supreme Court ruling preventing procurement of commercial food in its feeding programmes. The central government issued a ban on subsequent RUTF shipments and asked the UN agency to take back the remaining stocks of RUTF. The same minimalist, quick intervention thus came to be seen as a contravention of the Indian state's protocols and procedures. As per a Reuter's news report, quoting a senior Indian official, 'I can understand their sympathies, but sometimes emotions cannot sweep away the procedure and protocol involved' (Reuters 2009). Ultimately, the result of this saga led to a more nuanced approach by the humanitarian organisations, who opted instead to take on board considerations from the Indian stakeholders and agreed to a consensus-building measure and to 'Indianise' RUTF (Doyon 2011).

Tom Scott-Smith (2013) observes that with the fetishisation of humanitarian commodities comes a tendency to obfuscate the multiple forms of valuation of these commodities to focus on just one: biological valuation. The product should save lives, at a scale that is at once minimalistic and universal. Going by the account of MSF's own staff, however, the panoply of stakeholders who engaged in debates on how to articulate RUTF revealed a 'repertoire of values' being mobilised (Heuts and Mol 2013). While MSF staff were fixated on the use value of the product ('its ability to save lives of children'), civilian stakeholders were concerned that the introduction of RUTF would pave the way for a commodification of nutrition solutions, opening these up to the market. To some of them, RUTF also represented a move towards the individualisation of nutrition solutions. But there were again other ways of assessing the product: the first usages of RUTF by an Indian state government signalled another value attached to RUTF, namely its ability to quickly deliver visible, tangible results. In contrast, the central government viewed this introduction of RUTF as a contravention of its sovereignty, a symbol of the transgovernmental overreach of humanitarian organisations. These multiple interventions give a first glimpse of the diversity of values attached to RUTF by Indian stakeholders. We can now detail some of these views, in relation to the dual status of India as a manufacturing and consumption site for these products.

RUTF AS AN IDEA AND A PRODUCT — EVALUATIONS BY CIVILIAN STAKEHOLDERS

While the previous section foregrounds the entanglement of the various valuation practices upheld by stakeholders, the next three sections go into the detail of these valuation practices, dealing successively with civilian stakeholders, private industry, and regulatory authorities. Let us first look at how civilian stakeholders have engaged with RUTF.

Civil society activists and other allied groups

This category comprises activist groups such as the Right to Food Campaign, which was an 'informal network of individuals and organisations committed to the realisation of the right to food in India'. For the stakeholders in this category, RUTF was articulated as an idea that introduced technologisation and private interest into food security issues. The first argument is about the cause of acute malnutrition in India. SAM is recognised as an acute emergency due to its high case fatality rates, prompting rapid intervention. However, chronic undernutrition, such as stunting and low weight for the sufferer's age, require holistic interventions. Dasgupta et al. (2014) highlight that India faces chronic undernutrition (termed severe chronic malnutrition), with seasonal SAM episodes exacerbating the issue. The lower anthropometric values in India are linked to the Multi-Dimensional Poverty Index, indicating broader causes beyond civilian emergencies alone. This is reflected in India's lower SAM case fatality rates (1.2%) compared to WHO's global estimates (10–20%) in a study of over 2,600 children aged 6–18 months (Prost et al. 2019). Moreover, a large-scale trial using RUTFs showed lower weight gain rates in Indian children with SAM than in Africa, despite better community care and more frequent therapy intervals (Bhandari et al. 2016). Prasad, Holla, and Gupta (2009), members of the Working Group on Children Under Six[15], a group comprising paediatricians, nutritionists, and economists involved in nutrition and food security advocacy in India, noted that (at the time) the efficacy of RUTF had been demonstrated only in contexts of disaster and famine in a few African countries (Prasad et al. 2009; Working Group on Children under Six 2007).

RUTF is also articulated as a centralising force that concentrates resources, including centralisation of production models and centralisation of technological integration. Coming to the first category, the RUTF has been interrogated for the unique position it occupies as a food product that behaves like a pharmaceutical. This has variously been described as the medicalisation[16] and the pharmaceuticalisation[17] of food (Caremel and Issaley 2016). Critics of the prevalent model of production of RUTF in India point to the fact that centralised production models strip affected communities of the power to decide. One of my interviewees, a public health practitioner, pointed out that this does not constitute an 'anti-technology' approach. The real issue, in their opinion, was the level of technological integration, which does not take place at community level. They observe,

> if you need calorie-dense high protein food then that is something we can do with community and community can do on its own when we need to add something we have always done it by giving supplements, so if you need a micronutrient particular mix to be added, let that be added at the level of house or community and so on.

The argument put forward here is that intertwining food with 'technological' elements (in mineral mixes, for example) creates a need for a certain type of production model that takes power away from communities. This is not to be understood as an 'anti-private sector' stance, but the idea is to achieve a better distribution of value generated from such an industry, which can be done through cooperatives.

Academic and programmatic evaluators

Beyond the analysis of RUTF as a nutritional and industrial product, what about the claim made on its behalf that it constitutes a magic bullet? What evidence is available about its 'performance' on the ground? Its efficacy, acceptability, and efficiency have been observed by practitioners, academics, and paediatric doctors in India, who employ knowledge-making procedures similar to the ones employed to prove claims about RUTF in the first place, such as field trials and randomised trials. For example, one of the attractive qualities of industrially produced RUTF is its sterility and resistance to contamination. However, in field trials comparing different RUTF formulations in India, diarrhoea is reported to have occurred after people drank unclean water in intervention areas (Garg et al. 2018). This is echoed by an interviewee, who is a practitioner of public health:

> Somebody will put (RUTF) in spoon and take it out some and then keep it and then take another spoon, so all the sterility goes for … it goes for a big six [a colloquial cricket expression meaning some idea going to waste/ out of contention] and then you have to give a lot of water on top of RUTF because it's so dense and sweet (and) the water is not sterile anyway. Are you dealing with the water quality at household level?

Debates on the acceptability and appropriateness of RUTF are not new, as one single formulation is used across diverse cultures and geographies. In the proverbial 'humanitarian reason', taste is an afterthought to the more pressing need to save lives. A quotation from Micah Trapp's (2016) paper perfectly captures the idea: 'In the biopolitical domain of the refugee camp, food is a site of gustatory discipline'.

Against this reason, Indian stakeholders have generated evidence through field trials to sometimes argue for more locally tailored RUTF that aligns with people's taste preferences. Trials in India have compared locally produced RUTF (LRUTF) with commercially produced RUTF (CRUTF) and home-prepared foods. The results are mixed. CRUTF shows higher mean weight gain in children with SAM in some studies[18] (Dube et al. 2009). One large trial found higher recovery rates with LRUTF (Bhandari et al. 2016), though both RUTF types were less accepted than home-prepared foods (Dube et al. 2009) or LRUTF over CRUTF in some trials (Selvaraj et al. 2022). The costs of LRUTF and CRUTF are similar (Garg et al. 2018). Typically, RUTF is distributed with a clinical focus, ensuring strict adherence to the feeding regimen. Community-based programmes shift the responsibility of medical care to families, leading to the pharmaceuticalisation of food. Indian researchers emphasise RUTF's acceptability and taste by comparing it with LRUTF and energy-dense foods, thus repositioning RUTF as food, as one public health practitioner comments: 'food is something that we engage within our own kitchen, household, and in our own communities. And there is certain value in retaining rather than medical-izing food'. To summarize, civilian stakeholders articulate RUTF as an idea, a production model, and a product. The activist groups foreground questions of production models and centralisation of value addition/creation in the prevalent mode of RUTF procurement today. If we understand magic bullets (RUTF here) as single value objects, then asking these questions helps situate RUTF inside a web of complex sociopolitical interrelations falling within the evolutions of humanitarianism today. RUTF is not just a freestanding product; it also stands for its entire supply chain. The other group of academics and knowledge crea-tors examine its programmatic claims by generating evidence using controlled and field trials. Emphasising taste in food aid is often understood as a way for recipients of aid to resist the agnosticism of humanitarianism, where taste is an afterthought (Trapp 2016). Comparing the acceptability of standard RUTF with locally available high-density foods, they reignite questions about taste and divergent usages of RUTF.

Private sector, appropriation, and production politics

The previous section focuses on how civil society stakeholders articulated the valuation of RUTF, illuminating questions about the centralisation of value creation, production models and hygiene/taste. Carrying forward the logic of technical objects moving through different regimes of valuation, we now focus our attention to what Appadurai (1988) would call the commodity phase of

RUTF. This section presents a snapshot of the nascent RUTF industry in India and the role that it plays in transcontinental trade circuits. The section then puts forward the tensions created by the drop in domestic demand for RUTF in India to enquire what other values are attached to RUTF, and consequently, how they are leveraged to sustain business.

Since the beginning of large scale RUTF procurement by UNICEF, local producers in the Global South, especially Africa, have faced price disadvantages compared to offshore RUTF manufacturers due to taxes on raw material imports, low factory utilisation rates, and high business costs (Segrè et al. 2017). The patent for RUTF was owned by Nutriset, the French company that first started manufacturing RUTF under the brand name Plumpy'Nut. Under its PlumpyField arrangement, the company aimed to improve local manufacturers' access to funds, expertise, and materials to lower costs (Sanderson 2016). Despite this, by 2022, only 64% of the total RUTF volume was procured from local manufacturers, even though nearly 90% were located in programme countries[19].

UNICEF started first procuring RUTF in India from a PlumpyField network member in 2011, followed by an Indian subsidiary of a Norwegian company, Compact, in 2013. Since then, four more home-grown manufacturers have sprung up. From 2017 onwards, UNICEF reports show that RUTF procured from Asia (India accounts for five of the six suppliers from Asia) has a price advantage over other sources[20][21] In 2022, RUTF procured from Indian suppliers was cheaper on average (ranging from $38.18 to $42.86 per carton) than its African and European counterparts (ranging from to $41 to $54.75).[22] Indian producers have also managed to reduce the prices of their RUTF to a lower level than their European and American counterparts. Furthermore, Indian producers appear to be better protected against price fluctuations. For example, between 2021 and 2022, in a period marred by protracted global crises and increased raw material and freight costs,[23] compared to a major European producer (an increase of 13.77 euros per carton of RUTF corresponding to a 38.34% increase), Indian RUTF's price increase remained between 4.89% to 16.12% for all manufacturers (author calculations, based on data available from UNICEF[24]. Of the five suppliers in India, four are independent manufacturers while one is under the

PlumpyField franchise network. This is in contrast to Africa, which has six out of eleven manufacturers under the PlumpyField network and a lesser number of independent manufacturers. The network manufacturers produce the RUTF under the brand name Plumpy'Nut, whereas independent manufacturers are free to market their own unique brands of RUTF.

Despite this opportunity and a price advantage, the domestic demand for RUTF has dropped.[25] In 2020, industry representatives of RUTF manufacturers came together to form an industry group called CMAM (Community Based Management of Acute Malnutrition), with the stated aim of 'provid[ing] a platform for a national discourse on the eradication of malnutrition by catalysing the involvement of stakeholders and interested parties' (Hindu Businessline 2020). The formation of groups such as the CMAM association of India can be seen as a strategy to create a discourse around RUTF. For example, CMAM has published press notes advocating the use of RUTF to treat children with SAM in the aftermath of the COVID-19 pandemic[26][27]. An analysis of the press notes released by CMAM through various online media outlets reveals a focus on presenting the case for RUTF usage in India by citing its effectiveness in tackling SAM and extolling its alignment with the idea of domestic production (Atmanirbhar Bharat). While the powerful idea of RUTF appears to be appropriated to create a discourse on its usage, the focus on the idea of domestic production reveals one more way RUTF is valued as a critical, lifesaving product – hence the value in freeing it from issues faced by long transcontinental supply chains. Furthermore, some RUTF manufacturers demonstrate creativity in diversifying their product portfolio with other products in the consumable categories, by leveraging their manufacturing lines and certifications as assets.[28] In the end, private players in India move beyond being mere sites for RUTF production. They demonstrate this by creating industry groups, some of them branching out into diversified categories of free market products and creating RUTFs with unique brand identities.

In India, on one hand, dominant modes of humanitarian aid delivery are problematised and questioned, while on the other hand, private entities operate on these very production logics. RUTF thus circulates through these apparently opposed regimes of valuation, where at one point it is valued as a mere commodity, sometimes laced with moral/ethical rhetoric, travelling through global commodity chains. At some other points in India, it is understood as emblematic of the technologisation of food aid and humanitarian overreach, sometimes even infringing on the public health apparatus' purview to help children suffering from SAM. We will now focus on how these tensions are negotiated by the regulatory authorities in India.

Regulatory problems and boundary objects

RUTF has been variously articulated as a hybrid between food and medicine. The existence of such hybrids has often produced tensions for regulatory boundaries, which are set not just for administrative purposes, but also to regulate choice and risk (Frohlich 2021). These tensions are mostly resolved by boundary work, understood here as Gieryn (1983) has articulated: as a rhetorical device that uses creative language and philosophy to demarcate areas of influence where professional autonomy can be maintained, especially when such boundaries are not well defined. In this section, I look at the process of guideline setting for RUTF by the Codex Alimentarius[29] Committee on Nutrition and Foods for Special Dietary Uses (CCNFSDU), especially in the light of India's participation in these deliberations, to understand how regulatory bodies in India negotiate this tension and what valuation strategies are employed at this stage.

India's domestic policy for handling uncomplicated SAM reflects a preference for locally prepared energy-dense foods. NITI Ayog, India's apex public policy think tank, observes that special foods for children with SAM (without complications) are already being designed and provided in many Indian states based on local preferences. This recommendation aligns with the already existing Supplementary Nutrition Programme (SNP) and the Take-Home Ration (THR) provided by the Ministry of Women and Child Development's flagship initiative, the ICDS (Integrated Child Development Services)[30]. At the same time, India hosts sites for the production of RUTF for export.

In 2016, at the annual meeting of the CCNFSDU, a proposal was made to draft guidelines for defining the scope, contents, and labelling of RUTF. This proposal was made with the explicit intention of providing a framework to guide the global supply of RUTF. Subsequent editions of the annual CCNFSDU reports reveal that subsequent negotiations have included issues like labelling, ingredients, and marketing (CCNFSDU 2016, 2017, 2019).

Against this backdrop, it is interesting to look at the RUTF guidelines newly adopted by the Codex. At the 45th session of the Codex Alimentarius Commission, held in November 2022, a host of key changes were made to the RUTF specifications, and a globally accepted standard was finalised[31]. A new categorisation of Food for Special Medical Purposes (FSMP) and the new globally accepted ingredients list present a clear case of boundary work. This new standard particularly addresses some of the criticisms levelled at RUTF in the past. For example[32]:

> The new guidelines specify a protein quality score rather than limiting pro-tein sources. This allows producers and researchers to propose other high quality protein sources that could be less expensive and still well-liked by babies and infants.

This change seems to have come in response to the criticism regarding the acceptability of RUTF's formulation, which is milk- and peanut-based. At the same time, this change has the potential to encourage local production of RUTF, using locally acceptable protein sources. However, while this invalidates the earlier requirement of a specific type of protein (milk protein), it introduces a new measure of protein quality (PDCAAS) as the defining criterion for inclu-sion. This entails further technologisation of food ingredients that ultimately invites industrial production. As one NGO representative put it, '[The] more you put the norm, [the] more you professionalize'.

Similarly, while the reduced total sugar and fatty acid requirements seem to address earlier criticism of the high sugar content of RUTF (Bazzano et al. 2017), the technical guidelines regarding ingredients remain, further solidifying the necessity of an industrial production regime.

As per a FSSAI (Food Safety and Standards Authority of India) report available on its website, the Indian delegation at the CCNFSDU 2019 meet-ing intervened to add within the preamble of RUTF that it was only one 'of the *options for the dietary management of uncomplicated SAM (in 6 month-59 month age group)'* and advocated for deleting any text that may suggest its usage in other age groups. The report mentions that these suggestions were accepted. The final CODEX guideline reflects this:

> Ready-to-use therapeutic food (RUTF) is a WHO recommended option for the dietary management of children aged from 6 to 59 months with SAM without medical complications. **However, this does not preclude other dietary options including the use of locally-based foods.** RUTF is not for general retail sale. (FAO.org)

We can argue that the regulatory environment agreed upon by all the Codex delegations (including India) seems to favour industrial production and a global trade regime, while India's intervention mentioned above seems to create a space for potential alternatives to RUTF for treating SAM. The boundaries thus drawn, in my opinion, invite the conclusion that the regulatory body in India values RUTF in two different contexts. The participation in the global exercise seems to be indicative of the fact that RUTF is valued as an emergency commodity, especially one that can be traded, hence the need to formulate an internationally acceptable standard that does away with confusion and streamlines the global flows. Domestically, the limits of RUTF are defined to create a space for an indigenous way to tackle SAM (RUTF as one of the options, not marketed to any other age group). It is in this tightrope walk between industry interests and domestic public health priorities that the tensions mentioned in the beginning of the section appear to be resolved.

DISCUSSION AND CONCLUSION

This chapter contributes to the volume by illuminating an arguably atypical example of technoscientific globalisation from below and market-making. It interrogates the alternative ways and strategies that actors in the Global South employ to reimagine and rearticulate their position as recipients of humanitarian aid. These include indigenous knowledge production, appropriation of production models to create economic value, and the use of formal institutions and regulatory or judicial measures to challenge the seemingly unilateral imposition of the RUTF humanitarian programme. The chapter also highlights the ways in which a heterogeneity of valuation practices are presented by a humanitarian palliative, demonstrating a colourful panoply of stakeholders that engage with a powerful idea to their own respective ends. From the dynamic interplay between the central and state governments to regulatory complexities, the vocal opposition of civil society stakeholders, and the burgeoning private sector's capability to scale up RUTF production, the conventional categories of globalised humanitarianism have been questioned. These categories include the idea of an innovative North and a recipient South, and the notion that local institutions in the South are breaking down, necessitating humanitarian action. In the evolving discourse surrounding RUTF, there's a discernible trend suggesting a shift from technology-centric and morally driven discussions to considerations of unjust production models and the importance of public healthcare.

The introduction of new treatment models like Community-Based Management of Acute Malnutrition (CMAM) and the incorporation of RUTF as a specialised therapeutic food within these initiatives have catalysed a profound transformation in the discourse on malnutrition. Introduction of easy-to-use methodologies to identify malnutrition made the issue of SAM 'visible' (Scott-Smith 2013), which ultimately incentivised tangible interventions and consequently has driven huge funding increases for RUTF-based interventions. RUTF and a clinical articulation of malnutrition remains at the epicentre of the discussions on malnutrition.

The chapter ultimately tries to trace the magic bullet's trajectory, to see if it indeed followed a straight path. We see that in the case of humanitarian exchange with regard to RUTF and malnutrition, it is not quite so. The stakeholders in India deal with this 'finality' of a product ending up in the bodies of the disenfranchised malnourished in India with their own practices of valuation. It is in these examples that this chapter finds forms of resistance to the idea of a technological globalisation from above into matters of living.

ENDNOTES

1   This chapter presents a part of the PhD work carried out by the author at Université Paris Cité, Paris.

2   UNICEF https://www.unicef.org/child-alert/severe-wasting extracted on 18-09-2023

3   Rice, A. "The peanut solution." *The New York Times: Sunday Magazine* (2010): 36-40.

4   It is also important to mention that RUTF was distributed within a programmatic innovation called Community Based Management of Acute Malnutrition (CMAM) that constituted dividing SAM children into two categories of complicated cases needing hospitalisation and uncomplicated cases that could be treated at community level. This additional filter, along with simplified tools to detect SAM (mid upper arm circumference tapes), helped reduce the burden on hospitals and scale up relief efforts with minimal infrastructure requirement (interviews with NGO representative and nutritionist).

5   Nagarajan R (2017, September 12). No quick-fix solution: Don't use packaged food to fight malnutrition, says govt. *Times of India*. https://timesofindia.indiatimes.com/india/no-quick-fix-solution-dont-use-packaged-food-to-fight-malnutrition-says-govt/articleshow/60471282.cms extracted on 26-09-2024

6   Thacker T (2017, Novermber 9). PMO: States will take a call on ready-to-use food for children. *Mint.* https://www.livemint.com/Industry/Q59PA6oEQoUdGJZmVoPwLK/PMO-states-will-take-a-call-on-readytouse-food-for-childr.html extracted on 24-09-2024

7   Barnagarwala T (2017, October 10). Maharashtra: Nandurbar procures ready-to-use food despite Centre's notice. *Indian Express.* https://indianexpress.com/article/india/maharashtra-nandurbar-procures-ready-to-use-food-despite-centres-notice-4912959/ extracted on 01-07-2023

8   Johari A (2017 September 16). Maharashtra puts on hold its controversial plan to supply malnourished kids with therapeutic foods. *Scroll.in.* https://scroll.in/article/850746/maharashtra-puts-on-hold-its-controversial-plan-to-supply-malnourished-kids-with-therapeutic-foods extracted on 26-09-2024

9   Barnagarwala T (2019, January 10). Govt set to distribute therapeutic food to malnourished children,activists oppose. *Indian Express.* https://indianexpress.com/article/cities/mumbai/govt-set-to-distribute-therapeutic-food-to-malnourished-children-activists-oppose-5531116/ extracted on 01-07-2023

10   'The Codex Alimentarius is a collection of internationally adopted food standards and related texts presented in a uniform manner. These food standards and related texts aim at protecting consumers' health and ensuring fair practices in the food

trade. The publication of the Codex Alimentarius is intended to guide and promote the elaboration and establishment of definitions and requirements for foods to assist in their harmonisation and in doing so to facilitate international trade' (FAO.org).

11 UNICEF (2021) https://www.unicef.org/supply/media/7256/file/Ready-to-Use-Therapeutic-Food-Market-and-Supply-Update-March-2021.pdf extracted on 18-09-2023

12 UNICEF (2021) https://www.unicef.org/supply/media/8246/file/Supply-Annual-Report-2020.pdf. Accessed on 12-05-2025

13 UNICEF (2022) https://www.unicef.org/supply/media/12726/file/AnnexesSupply-Annual-Report-2021.pdf extracted on 01-07-2023

14 UNICEF (2021) https://www.unicef.org/supply/media/8246/file/Supply-Annual-Report-2020.pdf extracted on 12-05-2025

15 The Working Group on Children under Six is a joint effort of People's Health Movement – India and the Right to Food Campaign.

16 Medicalisation, as per Conrad (2005), refers to the fact of 'mak[ing] medical, any phenomena'.

17 Understood here in the way that Joao Beihl (2004) uses it and that Abraham (2010) defines it, as 'the process by which social, behavorial or bodily functions are treated or deemed to be treated by medical drugs'.

18 Shukla, A and Marathe, S. "The Malnutrition Market; Let Them Eat Paste." *Economic and Political Weekly,* Vol. 52, Issue No. 25-26, 24 Jun, 2017

19 UNICEF (2022) https://www.unicef.org/child-alert/severe-wasting extracted on 18-09-2023

20 UNICEF (2023) https://www.unicef.org/supply/media/17331/file/Ready-to-Use-Therapeutic-Food-Market-and-Supply-Update-May-2023.pdfExtracted on 27-05-2024 extracted on 18-09-2023

21 UNICEF also reports that until 2020, the RUTF procured from Asia as a share of total procurement was the highest, suggesting that the *quantum* of RUTF originating from India/Asia was quite high - https://www.unicef.org/supply/stories/new-codex-guidelines-pave-way-innovation-ready-use-therapeutic-food-rutf extracted on 18-09-2023.

22 Some figures were in euros. The conversion to dollars was done using a conversion factor of 1 euro = $1.0538, which is the average exchange rate in 2022, sourced from https://www.exchangerates.org.uk/EUR-USD-spot-exchange-rates-history-2022.html [accessed 11 June 2024].

23 UNICEF (2022) https://www.unicef.org/supply/media/19791/file/SupplyAnnualReport2022.pdf extracted on 07-06-2024 and UNICEF (2023) https://www.unicef.org/supply/media/17331/file/Ready-to-Use-Therapeutic-Food-Market-and-Supply-Update-May-2023.pdfExtracted on 27-05-2024

24 Price data from UNICEF (2022) https://www.unicef.org/supply/media/17916/file/Ready-to-use-therapeutic-food-price-data-2003-2022.pdf extracted on 11-06-2024

25 In 2023, as per a report jointly authored by NITI Ayog and UNICEF, RUTF was not used in the mentioned state-based CMAM programmes, all of which used energy-dense Take-Home Ration or other locally prepared products. https://shorturl.at/d4fqm accessed on 12-05-2025

26 Need for Accelerated Community Program to Counter SAM as Pandemic Ebbs: CMAM Association (2022, May 28). *Business Standard.* https://www.business-standard.com/content/press-releases-ani/need-for-accelerated-community-program-to-counter-sam-as-pandemic-ebbs-cmam-association-122032801426_1.html extracted on 24-09-2024

27 Hindustan Times, 'Therapeutic Food Makers Extend Support to Check Malnutrition During COVID Times', *Hindustan Times,* 26 May 2021 https://www.hindustantimes.com/brand-post/therapeutic-food-makers-extend-support-to-check-malnutrition-during-covid-times-101622036453217.html. extracted on 24-09-2024

28 One such private company markets itself as a supplier of white label peanut butter for firms in Europe. While it is unclear if this has actually translated into business, the same company producing RUTF for humanitarian needs in Africa and white label peanut butter for companies in Europe is an interesting idea to consider (https://www.nuflowerfoods.com/blogs/the-business-of-peanut-butter-in-europe-white-labelling-and-partnering-with-manufacturers-in-india/ and https://www.nuflowerfoods.com/factory-overview/ [accessed 7 June 2024]).

29 'The Codex Alimentarius is a collection of internationally adopted food standards and related texts presented in a uniform manner. These food standards and related texts aim at protecting consumers' health and ensuring fair practices in the food trade. The publication of the Codex Alimentarius is intended to guide and promote the elaboration and establishment of definitions and requirements for foods to assist in their harmonisation and in doing so to facilitate international trade' (FAO.org)

30 NITI Ayog (2023) https://shorturl.at/d4fqm accessed on 12-05-2025

31 UNICEF (2023) https://www.unicef.org/supply/media/17331/file/Ready-to-Use-Therapeutic-Food-Market-and-Supply-Update-May-2023.pdf Extracted on 27-05-2024

32 UNICEF (2022) https://www.unicef.org/supply/stories/new-codex-guidelines-pave-way-innovation-ready-use-therapeutic-food-rutf extracted on 18-09-2023

REFERENCES

Abraham, J., 'Pharmaceuticalization of Society in Context: Theoretical, Empirical and Health Dimensions', *Sociology,* 44.4 (2010): 603–22.

Anderson, W., 'Introduction: Postcolonial Technoscience', *Social Studies of Science*, 32.5–6 (2002): 643–58.

Appadurai, A., *The Social Life of Things: Commodities in Cultural Perspective* (Cambridge University Press, 1988).

Ashworth, A., 'Efficacy and Effectiveness of Community-Based Treatment of Severe Malnutrition', *Food and Nutrition Bulletin*, 27.3 Suppl (2006): S24–48.

Ashworth, A., S. Huttly, and S. Khanum, 'Controlled Trial of Three Approaches to the Treatment of Severe Malnutrition', *The Lancet*, 344.8939–40 (1994): 1728–32.

Bazzano, A. N., K. S. Potts, L. A. Bazzano, and J. B. Mason, 'The Life Course Implications of Ready to Use Therapeutic Food for Children in Low-Income Countries', *International Journal of Environmental Research and Public Health*, 14.4 (2017): 403.

Beesabathuni, K. N., and U. C. M. Natchu, 'Production and Distribution of a Therapeutic Nutritional Product for Severe Acute Malnutrition in India: Opportunities and Challenges', *Indian Pediatrics*, 47 (2010): 702–6.

Bhandari, N., and others, 'Efficacy of Three Feeding Regimens for Home-Based Management of Children with Uncomplicated Severe Acute Malnutrition: A Randomised Trial in India', *BMJ Global Health*, 1.4 (2016): e000144.

Biehl, J. G., 'The Activist State: Global Pharmaceuticals, AIDS, and Citizenship in Brazil', *Social Text*, 22.3 (2004): 105–32.

Black, R. E., and others, 'Maternal and Child Undernutrition: Global and Regional Exposures and Health Consequences', *The Lancet*, 371.9608 (2008): 243–60.

Briend, A., et al, 'Ready-to-Use Therapeutic Food for Treatment of Marasmus', *The Lancet*, 353.9166 (1999): 1767–68.

Briend, A., and Collins, S. 'Therapeutic nutrition for children with severe acute malnutrition: summary of African experience.' *Indian Pediatrics* 47 (2010): 655-659.

Burza, S., and others, 'Community-Based Management of Severe Acute Malnutrition in India: New Evidence from Bihar', *The American Journal of Clinical Nutrition*, 101.4 (2015): 847–59.

Caremel, J.-F., and N. Issaley, 'Des Cultures Alimentaires "Sous Régime d'Aide"? Négociations Autour des Aliments Thérapeutiques Prêts à l'Emploi (ATPE) et de la Malnutrition Infantile (Sahel)', *Anthropology of Food*, 11 (2016).

CCNFSDU, 'Report of the 41st Session of the Codex Committee on Nutrition and Foods for Special Dietary Uses' (2016) https://www.fao.org/fao-who-codexalimentarius/sh-proxy/en/?lnk=1&url=https%253A%252F%252Fworkspace.fao.org%252Fsites%252Fcodex%252FMeetings%252FCX-720-37%252FREP16_NFSDUe.pdf.

CCNFSDU, 'Report of the 42nd Session of the Codex Committee on Nutrition and Foods for Special Dietary Uses' (2017) https://www.fao.org/fao-who-codexalimentarius/sh-proxy/en/?lnk=1&url=https%253A%252F%252Fworkspace.fao.org%252Fsites%252Fcodex%252FMeetings%252FCX-720-38%252FReport%252FFINAL%252FREP17_NFSDUe.pdf.

CCNFSDU, 'Report of the 44th Session of the Codex Committee on Nutrition and Foods for Special Dietary Uses' (2019) https://www.fao.org/fao-who-codexalimentarius/sh-proxy/en/?lnk=1&url=https%253A%252F%252Fworkspace.fao.org%252Fsites%252Fcodex%252FMeetings%252FCX-720-40%252FREPORT%252FREP19_NFSDUe.pdf.

Collier SJ, Cross J, Redfield P, Street A. 'Preface: Little development devices/Humanitarian goods'. *Limn.* (2017) Nov 1;9

Ciliberto, M. A., et al, 'Comparison of Home-Based Therapy with Ready-to-Use Therapeutic Food with Standard Therapy in the Treatment of Malnourished Malawian Children: A Controlled, Clinical Effectiveness Trial', *The American Journal of Clinical Nutrition*, 81.4 (2005): 864–70.

Collins, S., and K. Sadler, 'Outpatient Care for Severely Malnourished Children in Emergency Relief Programmes: A Retrospective Cohort Study', *The Lancet*, 360.9348 (2002): 1824–30.

Conrad, P., 'The Shifting Engines of Medicalization', *Journal of Health and Social Behavior*, 46.1 (2005): 3–14.

Dahdah, M. Al, A. Kumar, and M. Quet, 'Empty Stocks and Loose Paper: Governing Access to Medicines through Informality in Northern India', *International Sociology*, 33.6 (2018): 778–95.

Dasgupta, R., D. Sinha, and V. Yumnam, 'Programmatic Response to Malnutrition in India: Room for More than One Elephant?' *Indian Pediatrics*, 51 (2014): 863–68.

Daudet, A., and C. Navarro-Colorado, 'Socio-Anthropological Aspects of Home Recovery from Severe Malnutrition', *Field Exchange*, 21 (2004): 23.

De Laet, M., and A. Mol, 'The Zimbabwe Bush Pump: Mechanics of a Fluid Technology', *Social Studies of Science*, 30.2 (2000): 225–63.

Der Geest, S., and others, 'The Urgency of Pharmaceutical Anthropology: A Multilevel Perspective', *Curare*, 34.1 (2011): 2.

Dhar, A., 'India's Vaccine Diplomacy', *India ChinaInstitute* (2021) https://www.indiachinainstitute.org/2021/03/04/indias_vaccine_diplomacy/.

Diop, E. H. I., and others, 'Comparison of the Efficacy of a Solid Ready-to-Use Food and a Liquid, Milk-Based Diet for the Rehabilitation of Severely Malnourished Children: A Randomized Trial', *The American Journal of Clinical Nutrition*, 78.2 (2003): 302–7.

Doyon, S., 'India: The Expert and the Militant', in A. Wilkins, ed., *Humanitarian Negotiations Revealed: The MSF Experience* (C. Hurst & Co, 2011), pp. 147–60.

Dube, B., and others, 'Comparison of Ready-to-Use Therapeutic Food with Cereal Legume-Based Khichri Among Malnourished Children', *Indian Pediatrics*, 46 (2009): 383–88.

Ecks, S., and S. Basu, 'The Unlicensed Lives of Antidepressants in India: Generic Drugs, Unqualified Practitioners, and Floating Prescriptions', *Transcultural Psychiatry*, 46.1 (2009): 86–106.

ENN, 'Comparison of the Efficacy of a Solid Ready-to-Use Food and a Liquid, Milk-Based Diet in Treating Severe Malnutrition', *Field Exchange*, 20 (2003): 4.

Fassin, D., 'Humanitarianism as a Politics of Life', *Public Culture*, 19.3 (2007): 499–520.

Fassin, D., *Humanitarian Reason: A Moral History of the Present* (University of California Press, 2011).

Fassin, D., 'This Is Not Medicalization', in M. Milhet, and G. Hunt, eds., *Drugs and Culture* (Routledge, 2016), pp. 107–16.

Frohlich, X., 'Humanitarianism as Medicalization', *International Journal of Applied Medical Sciences*, 22.2 (2021): 164.

FSSAI (Food and Safety Standards Authority of India), 'Brief Report of the 41st Codex Committee on Nutrition and Foods for Special Dietary Uses' (2020) https://www.fssai.gov.in/upload/uploadfiles/files/Brief_Report_41st_Codex_CCNFSDU_03_02_2020.pdf.

Garg, C. C., S. Mazumder, and others, 'Costing of Three Feeding Regimens for Home-Based Management of Children with Uncomplicated Severe Acute Malnutrition from a Randomised Trial in India', *BMJ Global Health*, 3 (2018): e000702. https://www.doi.org/10.1136/bmjgh-2017-000702.

Gera, T., 'Efficacy and Safety of Therapeutic Nutrition Products for Home-Based Therapeutic Nutrition for Severe Acute Malnutrition: A Systematic Review', *Indian Pediatrics*, 47 (2010): 709–18.

Gieryn, T. F., 'Boundary-Work and the Demarcation of Science from Non-Science: Strains and Interests in Professional Ideologies of Scientists', *American Sociological Review*, 48 (1983): 781–95.

Greiner, T., 'The Advantages, Disadvantages and Risks of Ready-to-Use Foods', *Breastfeeding Briefs*, 56.57 (2014): 1–22.

Grover, Z., and L. C. Ee, 'Protein Energy Malnutrition', *Pediatric Clinics*, 56.5 (2009): 1055–68.

Hardon, A., and Sanabria, E., 'Fluid Drugs: Revisiting the Anthropology of Pharmaceuticals', *Annual Review of Anthropology*, 46 (2017): 117–32.

Heuts, F, and Mol, A. 'What is a good tomato? A case of valuing in practice.' *Valuation Studies* 1, no. 2 (2013): 125-146.

Indian Express, 'Maharashtra Procures Ready-to-Use Food Despite Centre's Notice', *Indian Express*, 30 October 2017 https://indianexpress.com/article/india/maharashtra-nandurbar-procures-ready-to-use-food-despite-centres-notice-4912959/.

Indian Express, 'Govt Set to Distribute Therapeutic Food to Malnourished Children, Activists Oppose', *Indian Express*, 10 January 2019 https://indianexpress.com/article/cities/mumbai/govt-set-to-distribute-therapeutic-food-to-malnourished-children-activists-oppose-5531116/.

Jézéquel, J.-H., 'Staging a "Medical Coup"? Médecins Sans Frontières and the 2005 Food Crisis in Niger', *Medical Humanitarianism: Ethnographies of Practice* (2015): 119–35.

Khanum, S et al. "Controlled trial of three approaches to the treatment of severe malnutrition." *Lancet* vol. 344,8939-8940 (1994): 1728-32. doi:10.1016/s0140-6736(94)92885-1

Khara, T., and C. Dolan, *The Relationship Between Wasting and Stunting, Policy Programming and Research Implications* (USAID, 2018).

Komrska, J., L. R. Kopczak, and J. M. Swaminathan, 'When Supply Chains Save Lives', *Supply Chain Management Review*, 17.1 (2013): 42–49.

Kopytoff, I., 'The Cultural Biography of Things: Commoditization as Process', *The Social Life of Things: Commodities in Cultural Perspective*, 68 (1986): 70–73.

Lakoff, A., *Pharmaceutical Reason: Knowledge and Value in Global Psychiatry* (Cambridge University Press, 2006).

Lock, M. M., and V.-K. Nguyen, *An Anthropology of Biomedicine* (John Wiley & Sons, 2018).

Magone, C., M. Neuman, and F. Weissman, *Humanitarian Negotiations Revealed: The MSF Experience* (Oxford University Press, 2012).

Manary, M. J., and others, 'Home-Based Therapy for Severe Malnutrition with Ready-to-Use Food', *Archives of Disease in Childhood*, 89.6 (2004): 557–61.

Mason, J. B., and B. M. Margetts, 'Magic Bullets vs Community Action: The Trade-Offs Are Real', *World Nutrition*, 8.1 (2017): 5–25.

Mates, E., and K. Sadler, *Ready-to-Use Therapeutic Food (RUTF) Scoping Study* (Emergency Nutrition Network, 2020). https://www.ennonline.net/resource/ready-use-therapeutic-food-rutf-scoping-study

Ministry of Health and Family Welfare, *National Family Health Survey (NFHS-5) Phase II* (2021) https://main.mohfw.gov.in/sites/default/files/NFHS-5_Phase-II_0.pdf.

Ministry of Woman and Child Development, *National Nutrition Policy* (1993) https://wcd.nic.in/sites/default/files/nnp_0.pdf.

Ministry of Woman and Child Development, *RUTF Guidelines* (2017) https://wcd.nic.in/sites/default/files/RUTF.PDF.

Mint, 'PMO States Will Take a Call on Ready-to-Use Food for Children', *Mint* (2017) https://www.livemint.com/Industry/Q59PA6oEQoUdGJZmVoPwLK/PMO-states-will-take-a-call-on-readytouse-food-for-childr.html.

National Food Security Act (NFSA), *NFSA Act* (2013) https://nfsa.gov.in/portal/nfsa-act.

Ndekha, M. J., P. Ashorn, and A. Briend, 'Home-Based Therapy with Ready-to-Use Therapeutic Food Is of Benefit to Malnourished, HIV-Infected Malawian Children', *Acta Paediatrica*, 94.2 (2005): 222–25.

Patel, M. P., and others, 'Supplemental Feeding with Ready-to-Use Therapeutic Food in Malawian Children at Risk of Malnutrition', *Journal of Health, Population and Nutrition*, 23 (2005): 351–57.

Pordié, L., 'Pervious Drugs: Making the Pharmaceutical Object in Techno-Ayurveda', *Asian Medicine*, 9.1–2 (2014): 49–76.

Prasad, V., R. Holla, and A. Gupta, 'Should India use Commercially Produced Ready to Use Therapeutic Foods (RUTF) for Severe Acute Malnutrition (SAM)', *Social Medicine* 4.1 (2009): 52–55.

Prasad, V., D. Sinha, and S. Sridhar, 'Falling Between Two Stools: Operational Inconsistencies between ICDS and NRHM in the Management of Severe Malnutrition', *Indian Pediatrics* 49 (2012): 181.

Prost, A., and others, 'Mortality and Recovery Following Moderate and Severe Acute Malnutrition in Children Aged 6–18 Months in Rural Jharkhand and Odisha, Eastern India: A Cohort Study', *PLOS Medicine*, 16.10 (2019): e1002934.

Quet, M., 'Values in Motion: Anti-Counterfeiting Measures and the Securitization of Pharmaceutical Flows', *Journal of Cultural Economy*, 10.2 (2017): 150–62.

Redfield, P., 'Bioexpectations: Life Technologies as Humanitarian Goods', *Public Culture*, 24.1 (2012): 157–84.

Redfield, P., 'Fluid Technologies: The Bush Pump, the LifeStraw and Microworlds of Humanitarian Design', *Social Studies of Science*, 46.2 (2016): 159–83 https://doi.org/10.1177/0306312715620061.

Redfield, P., 'On Band-Aids and Magic Bullets', *Limn*, 9 (2017).

Reuters, 'India's Malnutrition Crisis: Therapeutic Food for Children', *Reuters*, 4 August 2009 https://www.reuters.com/article/idUSDEL499363/.

Ruxin, Joshua Nalibow. *Hunger, science and politics: FAO, WHO, and Unicef nutrition policies, 1945-1978.* (University of London, 1996).

Sanderson, J., 'Can Intellectual Property Help Feed the World? Intellectual Property, the PLUMPYFIELD® Network and a Sociological Imagination', in *The Intellectual Property and Food Project* (Routledge, 2016), pp. 145–74.

Sandige, H., and others, 'Home-Based Treatment of Malnourished Malawian Children with Locally Produced or Imported Ready-to-Use Food', *Journal of Pediatric Gastroenterology and Nutrition*, 39.2 (2004): 141–46.

Sathyamala, C., 'Nutrition as a Public Health Problem (1900–1947)', *ISS Working Paper Series/General Series*, 510 (2010): 1–30.

Saunik, S., and others, 'Safety, Tolerability, Efficacy and Logistics of Administration of Three Types of Therapeutic Feeds to Children with Severe Acute Malnutrition (SAM)', *International Journal of Nutrition*, 3.1 (2018): 10–15 https://doi.org/10.14302/issn.2379-7835.ijn-18-2262.

Scott-Smith, T., 'The Fetishism of Humanitarian Objects and the Management of Malnutrition in Emergencies', *Third World Quarterly*, 34.5 (2013): 913–28.

Scott-Smith, T., 'Humanitarian Neophilia: The "Innovation Turn" and Its Implications', *Third World Quarterly*, 37.12 (2016): 2229–51.

Scott-Smith, T., 'Sticky Technologies: Plumpy'nut®, Emergency Feeding and the Viscosity of Humanitarian Design', *Social Studies of Science*, 48.1 (2018): 3–24 https://doi.org/10.1177/0306312717747418.

Scott-Smith, T., *On an Empty Stomach: Two Hundred Years of Hunger Relief* (Cornell University Press, 2020).

Scroll.in, 'Maharashtra Puts on Hold Its Controversial Plan to Supply Malnourished Kids with Therapeutic Foods', *Scroll.in*, 16 September 2017 https://scroll.in/article/850746/maharashtra-puts-on-hold-its-controversial-plan-to-supply-malnourished-kids-with-therapeutic-foods.

Segrè, J., G. Liu, and J. Komrska, 'Local Versus Offshore Production of Ready-to-Use Therapeutic Foods and Small Quantity Lipid-Based Nutrient Supplements', *Maternal & Child Nutrition*, 13.4 (2017): e12376.

Selvaraj, K., R. S. Mamidi, R. Peter, and B. Kulkarni, 'Acceptability of Locally Produced Ready-to-Use Therapeutic Food (RUTF) in Malnourished Children: A Randomized, Double-Blind, Crossover Study', *Indian Journal of Pediatrics*, 89.11 (2022): 1066–72.

Strebhardt, K., and A. Ullrich, 'Paul Ehrlich's Magic Bullet Concept: 100 Years of Progress', *Nature Reviews Cancer*, 8.6 (2008): 473–80.

Street, A., 'Food as Pharma: Marketing Nutraceuticals to India's Rural Poor', *Critical Public Health*, 25.3 (2015): 361–72.

The Hindu Businessline, 'CMAM Association Formed to Address Severe Malnutrition in India Among Children', *The Hindu Businessline*, 26 May 2020 https://www.thehindubusinessline.com/news/cmam-association-formed-to-address-severe-malnutrition-in-india-among-children/article31676391.ece

Times of India, 'No Quick Fix Solution, Don't Use Packaged Food to Fight Malnutrition, Says Govt', *Times of India*, 12 September 2017 https://timesofindia.

indiatimes.com/india/no-quick-fix-solution-dont-use-packaged-food-to-fight-malnutrition-says-govt/articleshow/60471282.cms.

Trapp, M. M. 'You-Will-Kill-Me-Beans: Taste and the Politics of Necessity in Humanitarian Aid', *Cultural Anthropology*, 31.3 (2016): 412–37.

UNICEF (United Nations Children's Fund), 'Ready-to-Use Therapeutic Food Market and Supply Update – March 2021', (2021), https://www.unicef.org/supply/media/7256/file/Ready-to-Use-Therapeutic-Food-Market-and-Supply-Update-March-2021.pdf.

Whyte, S. R., S. der Geest, and A. Hardon, *Social Lives of Medicines* (Cambridge University Press, 2002).

Working Group on Children under Six, 'Strategies for Children under Six', *Economic and Political Weekly* 42.52 (2007): 87–101 http://www.jstor.org/stable/4027713.

World Food Programme (WFP). *Food assistance for children suffering from severe malnutrition: A toolkit for action* (WFP, 2020) https://www.wfp.org/publications/food-assistance-severe-malnutrition.

World Health Organization (WHO). 'Levels and trends in child malnutrition', UNICEF, WHO, and World Bank Group Joint Child Malnutrition Estimates (2021) https://www.who.int/publications/i/item/jme-2021.

Ziegler, T. R., and T. D. Wilkins, 'Nutritional Management of Severe Acute Malnutrition: The Role of Teady-To-Use Therapeutic Foods', *The American Journal of Clinical Nutrition*, 100.3 (2014): 573–81 https://doi.org/10.3945/ajcn.113.080586.

GLIMPSES OF THE TECHNOGLOB COLLECTIVE

MATTERING PRESS TITLES

The Ethnographic Case: Second Edition
EDITED BY EMILY YATES-DOERR AND CHRISTINE LABUSKI

Democratic Situations
EDITED BY ANDREAS BIRKBAK AND IRINA PAPAZU

Concealing for Freedom: The Making of Encryption, Secure Messaging and Digital Liberties
KSENIA ERMOSHINA AND FRANCESCA MUSIANI

Engineering the Climate: Science Politics, and Visions of Control
JULIA SCHUBERT

Environmental Alterities
EDITED BY CRISTÓBAL BONELLI AND ANTONIA WALFORD

With Microbes
EDITED BY CHARLOTTE BRIVES, MATTHÄUS REST AND SALLA SARIOLA

Energy Worlds in Experiment
EDITED BY JAMES MAGUIRE, LAURA WATTS, AND BRITT ROSS WINTHEREIK

Boxes: A Field Guide
EDITED BY SUSANNE BAUER, MARTINA SCHLÜNDER AND MARIA RENTETZI

An Anthropology of Common Ground: Awkward Encounters in Heritage Work
NATHALIA SOFIE BRICHET

Ghost-Managed Medicine: Big Pharma's Invisible Hands
SERGIO SISMONDO

Inventing the Social
EDITED BY NOORTJE MARRES, MICHAEL GUGGENHEIM AND ALEX WILKIE

Energy Babble
ANDY BOUCHER, BILL GAVER, TOBIE KERRIDGE, MIKE MICHAEL, LILIANA OVALLE, MATTHEW PLUMMER-FERNANDEZ, AND ALEX WILKIE

The Ethnographic Case
EDITED BY EMILY YATES-DOERR AND CHRISTINE LABUSKI

On Curiosity: The Art of Market Seduction
FRANCK COCHOY

Practising Comparison: Logics, Relations, Collaborations
EDITED BY JOE DEVILLE, MICHAEL GUGGENHEIM, AND ZUZANA HRDLIČKOVÁ

Modes of Knowing: Resources from the Baroque
EDITED BY JOHN LAW AND EVELYN RUPPERT

Imagining Classrooms: Stories of Children, Teaching and Ethnography
VICKI MACKNIGHT

www.ingramcontent.com/pod-product-compliance
Lightning Source LLC
Chambersburg PA
CBHW061301220326
41599CB00028B/5738